URBAN
REVITALIZATION

URBAN REVITALIZATION

Policies and Programs

Fritz W. Wagner
Timothy E. Joder
Anthony J. Mumphrey Jr.

editors

SAGE Publications
International Educational and Professional Publisher
Thousand Oaks   London   New Delhi

For information address:

SAGE Publications Ltd
1 Oliver's Yard
55 City Road
London EC1Y 1SP

SAGE Publications Inc.
2455 Teller Road
Thousand Oaks, California 91320

SAGE Publications India Pvt Ltd
B-42, Panchsheel Enclave
Post Box 4109
New Delhi 110 017

Printed in the United States of America

Library of Congress Cataloging-in-Publication Data

Main entry under title:

Urban revitalization : policies and programs / edited by Fritz W.
    Wagner, Timothy E. Joder, Anthony J. Mumphrey Jr.
        p. cm.
    Includes bibliographical references and indexes.
    ISBN 0-8039-5869-2 (cl.) — ISBN 0-8039-5870-6 (pb)
        1. Urban renewal—United States—Case studies.   2. Inner cities—
United States—Case studies.   3. Urban policy—United States—Case
studies. I. Wagner, Fritz W.   II. Joder, Timothy E.
III. Mumphrey, Anthony J.
HT175.U743   1995
307.3'416'0973—dc20   94-42555

This book is printed on acid-free paper.

95 96 97 98 99 10 9 8 7 6 5 4 3 2 1

Production Editor: Yvonne Könneker
Ventura Typesetter: Danielle Dillahunt

Contents

Foreword

We have endured a seemingly perpetual urban "crisis" now for at least the past 30 years. It was a crisis first perceived by the light of the fires that gutted numerous cities in the mid-1960s and one that sparked an initial wave of public concern. Erupting at the height of the era of "urban renewal," the riots mocked our earliest efforts to arrest the decay and social disorder that seemed endemic to older American cities after World War II.

We have since become, moreover, a suburban nation. Political currents similarly shifted, and attention paid to urban cores decreased dramatically in the 1970s and 1980s. The shift to a postindustrial service economy, the mobility of capital in a dynamic global marketplace, and the aging of urban infrastructures combined with racial and demographic trends to isolate the poorest populations, those with the least resources to cope with changing realities. Consequently, as minority populations have increased and, indeed, come to dominate many of our cities, the problems associated with the long-term decline of urban

America have deepened. Yet our cities remain. And their health remains essential to the nation's well-being.

It is a propitious moment to take a fresh look at what has been done for—and to—our central cities. The 1990s offer some hope for change. New political breezes are blowing (even if they are swirling and not definitively charted in any single direction), and cities have apparently crept back onto our national agenda. There seems to be, at any rate, a welcome disposition to do more than tell city dwellers to "vote with their feet."

The National Center for the Revitalization of Central Cities (NCRCC) has been charged with the task of assessing recent urban policy. That is a good place to start. In coming to grips with what has been tried, what has worked, and, clearly, what has not, the NCRCC can begin a process of reappraisal that is fundamental to any future progress.

In that sense, the case studies presented here represent a first step. The historical focus on central business district and economic redevelopment is examined on its own terms, in light of local successes and failures, and—at least by implication—we are invited to assess the adequacy of that approach in view of present conditions. Regional cooperation, the significance of human investment, the role of local elites, the reciprocal influences of planning and politics, and the relative weights to be assigned public and private actors are all encountered here. Especially important is the need to address and define the character of a potentially new federal-city relationship. If the plight of the inner-city poor themselves and policies directly affecting their life chances await further exploration, such future studies can benefit from this reconnaissance of the downtown-oriented programs that have previously claimed so much of our attention and treasure.

MARC H. MORIAL
Mayor of the City of New Orleans

ARNOLD R. HIRSCH
Professor of History, University of New Orleans

Introduction

Urban Revitalization Strategies for America's Central-Cities Matrix

In November 1991, urban researchers from several prominent U.S. universities convened in New Orleans at the invitation of the newly formed National Center for the Revitalization of Central Cities (NCRCC). Seeded by a $500,000 special-purpose grant through the U.S. Department of Housing and Urban Development, the NCRCC represented a unique undertaking on the part of the federal government. For the first time, funds were being devoted to direct outside consultation from the academic community on federal urban policy and program development.

The timing of this enterprise was doubly auspicious. In the spring of 1992 the nation's consciousness of urban ills was raised to a level not seen since Los Angeles's infamous Watts riots of 1965, as that city again exploded in violence. The economic and social conditions of inner cities throughout the nation came into sharp focus. At the same time, with the prospect of a new presidential administration on the horizon, the federal government's urban policies and programs became objects of scrutiny. Unfortunately, our collective understanding of the fixative process remained fuzzy.

With the advent of urban renewal in 1949, and for the succeeding 45 years, the federal government has taken a proactive role in urban development and revitalization, principally through a plethora of grant programs involving physical development, social transfer payments, and economic development initiatives ranging from job training to business loans. Their colloquial references are familiar to us: UDAG, CDBG, Model Cities; AFDC, Head Start, Section 8; Job Corps, EDA and FHA subsidies, for example. Many of these programs have been studied extensively; others, such as CDBG, are only now undergoing formal, intense, nationwide evaluation.

Regardless of the successes and failures attendant to individual federal programs and specific grants, the fact remains that the federal government has directed vast sums toward urban development and redevelopment, and yet this action has apparently failed to arrest the declines evident in many of our central cities. The charge to the NCRCC researchers assembled at the University of New Orleans that fall of 1991 was to study central-city revitalization programs in major U.S. cities, to analyze the successes and failures of various revitalization strategies, and to develop policy recommendations that would provide guidance to local communities and the federal government regarding the implementation of successful central-city revitalization programs.

The endeavor's inherent difficulty stemmed from the multiplicity of policies, programs, socioeconomic circumstances, projects, and financing tools that might be present in any revitalization effort. Quite simply, there was no way to control for, or estimate the impact of, all these various factors using the envisioned case study approach. Ultimately, the group concluded that the case studies should proceed on a variety of fronts using a generalized framework of (a) city context, covering government and administrative structures and socioeconomic conditions; (b) policy, program, and project descriptions; (c) impact assessment, using methodologies appropriate to the particular case; and (d) policy implications. All of the six case studies in this book, condensed from the originals, are presented in this format.

The investigators focused on seven prominent U.S. cities: Atlanta, Georgia; Baltimore, Maryland; Fort Worth, Texas; Minneapolis, Minnesota; New Orleans, Louisiana; New York, New York; and Portland, Oregon. These cities were chosen for several reasons, not all of them scientific. First, they represented a geographic blend of eastern and western, northern and southern cities. Their populations varied from more than 10 million to fewer than 500,000, and they include older and newer cities. Finally, as a matter of budget and logistics, they were familiar and/or convenient to the six NCRCC affiliate universities— City University of New York, University of Maryland, Rutgers University, Georgia Institute of Technology, University of Texas at Arlington, and, of course, the University of New Orleans, site of the NCRCC. The reader will readily discern that each of the case studies has a different central theme. Arthur Nelson of Georgia Tech compares development patterns in Atlanta and Portland, focusing on the importance of regional growth management to central-city revitalization. University of Maryland researchers David Imbroscio, Marion Orr, Timothy Ross, and Clarence Stone concentrate on the human element—Baltimore's efforts to revitalize the city through human investment programs. Edward Rogowsky, Ronald Berkman, Elizabeth Strom, and Anthony Maniscalco of the City University of New York examine Community Development Block Grant-related economic development efforts in Brooklyn and Queens, the core of the city's "outer borough development strategy." Mickey Lauria, Robert Whelan, and Alma Young of the University of New Orleans discuss the role of planning in revitalization efforts in that city, which they studied with assistance from the local Bureau of Governmental Research. Fort Worth's special districting approach to large-scale urban physical and economic redevelopment is chronicled by University of Texas at Arlington researchers Elise Bright, Richard Cole, and Sherman Wyman. Alex Schwartz at the New School for Social Research in New York (formerly with Rutgers University) examines the many downtown redevelopment projects in Minneapolis, as well as that city's efforts to channel revenue generated from downtown projects into neighborhood redevelopment.

Thematic considerations aside, the case studies presented in this volume constitute a broad survey of revitalization strategies in use over the past 15 to 20 years in major U.S. cities. Focusing on central-city revitalization efforts, the studies include examples of residential, commercial, and industrial revitalization; large, multiuse projects; inner-city neighborhood efforts; and more traditional downtown development.

Various revitalization strategies emphasizing physical development tools for individual projects, regional growth management, human investment, public-private partnership efforts, and tax and financial incentives were used in each of the cities studied by the NCRCC. The reader will note the following:

1. *The use of public investment as a catalyst* to spur private investment in a particular redevelopment project.

2. *The creation of downtown tax districts* as a tool, not only to raise funds to carry out particular projects but also as a device to manage and promote the city's core.

Strategy	Atlanta	Portland	New York	New Orleans	Fort Worth	Baltimore	Minneapolis
Public investment as catalyst	X	X	X	X	X	X	X
Creation of downtown tax districts		X	X	X	X		X
Private nonprofit development corporations	X		X			X	
Human investment strategies						X	
Commitment to planning as basis for action		X		X	X	X	X
Utilization of tax increment financing		X					X
Regional planning and governance		X					X
Creation of powerful development agencies		X	X				X
Extensive use of federal and state funds	X	X	X	X	X	X	X
Downtown development revenue to neighborhoods	X						X
Utilization of public-private partnerships	X	X	X	X	X	X	X
Transit as catalyst for downtown renewal		X					
Creation of urban growth limits		X					
Utilizing traumatic events	X	X	X				

Figure A.1 Revitalization Strategies Matrix

3. *The use of private, nonprofit development corporations.* A number of the case study cities found that complex and highly entrepreneurial projects could not be managed by traditional city agencies. They therefore created private, nonprofit entities to oversee project development.

4. *Human investment strategies.* Several cities, as part of their revitalization efforts, had components of those efforts deal directly with investment in human capital. Job training, education enhancement, career counseling, and other programs were targeted at the unemployed or underemployed in order to give those individuals the skills necessary to get and retain jobs in an increasingly service-oriented economy. Baltimore has made this human investment strategy the centerpiece of all its planning and revitalization efforts.

5. *A strong commitment to planning.* Many of the case studies show that successful projects had well-thought-out plans at their bases.

6. *Tax increment financing.* Several cities made extensive and productive use of the relatively unusual financial tool of tax increment financing. This allowed a number of cities to accumulate funds for public improvements that made large, mixed-use development possible by capturing the increased property values caused by the development.

7. *Regionwide managed growth.* The Portland and Minneapolis studies demonstrate that the respective success and failure of these cities' downtown revitalization efforts were tied directly to the effectiveness of managing the growth of their entire metropolitan areas and in carrying out specific programs on a regionwide basis. Programs of mass transit and solid waste disposal, regional tax equity programs, and regional growth boundary administration all contributed directly to strong central-city development efforts.

8. *Creation of powerful development agencies.* The case studies reveal that one very successful strategy is the creation of strong economic development agencies that have the independence, power, and financial resources necessary to move difficult projects through the development process. Although examples of the abuse of such power are evident, the value of having a competent and powerful development agency is also apparent.

9. *Use of federal and state development programs and funds.* All the cities in the case studies made extensive use of these development programs and funds. In fact, with the withdrawal of federal support for revitalization projects during the past decade, a number of revitalization projects were cut back or significantly delayed. The case studies reveal that federal support programs are definitely needed and utilized.

10. *Channeling revenue from successful downtown projects to neighborhoods.* This strategy, although relatively rare, today has great potential to rally support of neighborhoods for downtown projects, because the neighborhoods receive direct benefits from downtown development.

11. *Public-private partnerships.* Virtually all of the successful revitalization projects featured an emphasis on creating partnerships between the public and private sectors. Perhaps no other single strategy has been as critical to the success of redevelopment projects.

12. *Utilization of transit improvements as a catalyst for downtown renewal and private investment.* Whether it was building a transit station in a development project in New York, or building a regional transit system in Portland to direct growth, mass transit improvements have been shown to be a significant part of many revitalization strategies.

13. *Urban growth boundaries.* Although still relatively rare, urban growth boundaries have been shown to be a significant factor in the success of several city and regionwide revitalization efforts. Preliminary evidence from the case studies shows that where there is a regional consensus about growth, central-city revitalization efforts may have a greater chance for success.

14. *Cities that took advantage of potentially negative situations.* Although hardly a strategy to promote, taking a traumatic event and turning it into a catalyst for action has proven successful in some cities. Whether the threat of a downtown department store leaving or the potential for the relocation of an urban college, such an event, if properly acted on, can lead to successful revitalization projects.

There are other revitalization strategies that have been particular to only one or two of the case study cities, but the strategies described above are the ones most in use. They represent a wide variety of efforts, but each has shown itself a successful strategy, either alone or, most often, in partnership with a variety of strategies and programs.

The following chapters give examples of these strategies as well as some that have not been deemed successful, providing valuable learning tools for those who would seek to make America's central cities once again vibrant, growing centers of human activity. Our concluding essay coalesces the collective case study findings, describes key factors that universally contribute to revitalization successes and failures, and outlines six major federal policy recommendations that the contributors to this volume all agree are critical to an intelligent, coordinated federal response to urban problems.

FRITZ W. WAGNER
TIMOTHY E. JODER
ANTHONY J. MUMPHREY JR.

1

Regional Growth Management and Central-City Vitality

Comparing Development Patterns in Atlanta, Georgia, and Portland, Oregon

ARTHUR C. NELSON
with
JEFFREY H. MILGROOM

In this chapter, we offer the proposition that effective central-city revitalization cannot occur in the absence of effective management of regional development patterns. Efforts to effect revitalization through other means will fall short because of centripetal development inducements. We test this proposition by comparing the development patterns of Portland, Oregon, which has experienced the nation's highest level of regional development management, and Atlanta, Georgia, which represents the more normal approach to regional development management pursued in the United States. We compare Atlanta and Portland along several economic, social, and government dimensions, and then review major revitalization efforts employed by both central cities. We then discuss how regional urban growth management policies applied within each metropolitan area help explain emerging urban forms.

Profiling Atlanta and Portland

With some exceptions, the central cities of Atlanta and Portland are surprisingly similar in terms of history, economy, and socioeconomic factors. They are different in terms of government structure and regional development planning contexts.

LOCATION

The Atlanta metropolitan area, composed of 18 counties in 1990, is situated southeast of the Blue Ridge Mountains in north-central Georgia. The central county within which Atlanta is predominantly situated is Fulton. The suburban counties are Clayton, Cobb, DeKalb, and Gwinnett, and the exurban counties are Barrow, Butts, Cherokee, Coweta, Douglas, Fayette, Forsyth, Henry, Newton, Paulding, Rockdale, Spalding, and Walton. Population trends since 1960 are reported in Table 1.1 with respect to central-city and regional (metropolitan statistical area) growth.

The Portland metropolitan area, which comprises five counties, is situated at the confluence of the Willamette River and the Columbia River. The central county within which Portland is predominantly situated is Multnomah. The suburban counties are Clackamas and Washington on the Oregon side of the Columbia River, and Clark County on the Washington side. Yamhill County is the sole exurban county. However, because of very large land areas, much of the suburban counties and the eastern quarter of the central county can also be considered exurban, except that populations in the exurban portions of those counties are quite small. Population trends since 1960 are reported in Table 1.1 with respect to central-city and regional (metropolitan statistical area) growth.

HISTORY

Atlanta began in 1837 as the terminus of the Western and Atlantic Railroad; it was known as Terminus. The town was renamed Marthasville in 1845, and then again renamed Atlanta, after the Atlantic Railroad, in the same year by railroad engineer J. Edgar Thompson. It was incorporated in 1847. During the Civil War, Atlanta served as a Confed-

TABLE 1.1 Population Growth and Distribution, Atlanta and Portland
Central Cities and Metropolitan Areas, 1960-1990

Characteristic	Atlanta Central City	Atlanta Metropolitan Area	Portland Central City	Portland Metropolitan Area
1960 population	487,455	1,247,829	372,676	854,375
1970 population	496,973	1,684,200	382,619	1,047,343
1980 population	425,022	2,138,143	366,383	1,297,985
1990 population	394,017	2,833,511	437,319	1,477,895
Change 1960-1990	(93,438)	1,585,682	64,643	623,520
% change 1960-1990	-19.17	127.08	17.35	72.98
Regional growth share (%)	-5.89	100.00	10.37	100.00

SOURCE: U.S. Bureau of the Census (1991). Figures adjusted to include all metropolitan statistical area counties in 1990 for all censuses prior to 1990.

erate supply depot. In 1864, Union General William T. Sherman captured and burned Atlanta. Today, the seal of Atlanta is emblazoned with *Resurgens,* in tribute to its resurrection since that time. Until the early 1960s, Atlanta was in direct competition with Birmingham, Alabama, for regional prominence. The city has since grown into the South's dominant economic, commercial, and governmental center. In 1991, the city was awarded sponsorship of the 1996 summer Olympic Games. Since the 1960s, Atlanta has championed civil rights and racial equality.

Portland began as a port for shipping timber and agricultural products and importing finished goods in the late 1840s. It was incorporated in 1851 and named Portland because Francis W. Pettygrove of Portland, Maine, won a coin toss to determine the port's name over Asa L. Lovejoy of Boston, Massachusetts. Between 1883 and 1910, when Portland was connected by rail to Chicago and other points east, Portland was second only to San Francisco as the West Coast's largest city. In 1933, the "Tillamook burn" destroyed much of the timberland accessible to Portland, and the city lost much of its economic base. During World War II, more than 100,000 people moved into the city to build ships for the Japanese theater, and many tens of thousands stayed after the war. In the 1970s and 1980s, the city undertook several major downtown revitalization efforts. During the same time, Oregon adopted sweeping land use planning laws that preserved farmland and forest land around Portland for resource uses and directed urban development into Portland and its nearby suburbs.

ECONOMY AND EMPLOYMENT

More than 3,000 manufacturing firms are located in the Atlanta metropolitan area, employing about 15% of the labor force. The leading manufacturing industries are aircraft, automobile assembly, chemicals, communications equipment, textiles, metal products, and soft drinks. More than 3,000 wholesale firms make the city the South's dominant commercial center. Many of the world's wholesale transactions occur in several specialized trade-marts that are located downtown. Since 1960, however, the city of Atlanta has lost manufacturing employment while the metropolitan area has gained, a trend reported in Table 1.2. This is consistent with post-World War II trends, wherein manufacturing has become suburbanized. The total civilian labor force has barely risen in Atlanta since 1960, but it has increased substantially in the region. Those statistics are reported in Table 1.3.

Although it began as a center for shipping raw products and importing finished products, Portland has become among the nation's most diversified metropolitan economies, second only to Chicago in indices of diversity. More than 2,800 manufacturing plants are located in the Portland metropolitan area, and about one-eighth of the workforce is engaged in manufacturing. The leading manufacturing industry is metal processing, followed by electrical equipment, food products, lumber and wood products, paper, and transportation equipment. The Port of Portland, a statutorily authorized economic development district, handles more lumber and agricultural cargo than any other West Coast port. Since 1960, however, Portland has lost manufacturing employment while the metropolitan area has gained, a trend reported in Table 1.2. This is consistent with post-World War II trends, wherein manufacturing has become suburbanized. On the other hand, the total civilian labor force has grown substantially in both Portland and the region, as seen in Table 1.3.

SOCIOECONOMIC FACTORS

Relative to their metropolitan areas, Atlanta's and Portland's median family incomes have fallen, while the percentages of their populations that are black have risen. On the other hand, Table 1.4 shows that while Atlanta's regional share of families in poverty has remained fairly constant between 1979 and 1989, Portland's regional share of families in poverty has fallen.

TABLE 1.2 Manufacturing Employment, Atlanta and Portland Central Cities and Metropolitan Areas, 1967-1987

Characteristic	Atlanta Central City	Atlanta Metropolitan Area	Portland Central City	Portland Metropolitan Area
1967 employment	54,000	141,365	41,000	82,000
1972 employment	47,800	146,900	40,300	88,900
1977 employment	39,900	144,400	42,700	104,000
1982 employment	41,200	165,600	35,200	106,100
1987 employment	40,300	198,800	32,700	108,891
Change 1967-1987	(13,700)	57,435	(8,300)	26,891
% change 1967-1987	−28.66	39.10	−20.60	30.25
Regional growth share (%)	−23.85	100.00	−30.87	100.00

SOURCE: U.S. Bureau of the Census (1991). Figures adjusted to include all metropolitan statistical area counties in 1990 for all censuses prior to 1990.

TABLE 1.3 Civilian Labor Force, Atlanta and Portland Central Cities and Metropolitan Areas, 1960-1989

Characteristic	Atlanta Central City	Atlanta Metropolitan Area	Portland Central City	Portland Metropolitan Area
1960 civilian labor force	197,327	480,805	149,942	322,072
1970 civilian labor force	217,791	726,204	167,287	441,362
1980 civilian labor force	190,183	1,094,439	197,505	644,163
1989 civilian labor force	227,100	1,512,500	211,100	798,000
Change 1960-1989	29,773	1,031,695	61,158	475,928
% change 1960-1989	15.09	214.58	40.79	147.77
Regional growth share (%)	2.89	100.00	12.85	100.00

SOURCE: U.S. Bureau of the Census (1991). Figures adjusted to include all metropolitan statistical area counties in 1990 for all censuses prior to 1990.

GOVERNMENT

Atlanta has a mayor-council form of government composed of a strong mayor and a council president who are elected at-large and 18 council members elected from districts. The major departments of the city are administered by commissioners appointed by the mayor and approved by the council.

TABLE 1.4 Families Under Poverty, Atlanta and Portland Central Cities and Metropolitan Areas, 1969-1979

Characteristic	Atlanta Central City	Atlanta Metropolitan Area	Portland Central City	Portland Metropolitan Area
1979 families in poverty	18,973	44,538	7,851	19,097
1979 regional share	42.60%	100.00%	41.11%	100.00%
1989 families in poverty	23,152	55,120	11,628	30,266
1989 regional share	42.00%	100.00%	38.42%	100.00%

SOURCE: U.S. Bureau of the Census (1991). Figures adjusted to include all metropolitan statistical area counties in 1990 for all censuses prior to 1990.

Portland has a commission form of government composed of a mayor and four commissioners, all of whom are elected at-large. Portland is the largest city in the United States that does not elect its council members by district. Each of the five members of the city council heads a department. Although the mayor assigns departments and has important administrative powers, the mayor is not a strong executive.

REGIONAL DEVELOPMENT PLANNING CONTEXT

Atlanta is a member of the Atlanta Regional Commission (ARC), a council of governments that has special statutory powers to coordinate regional planning, social services, and transportation investments. The ARC has no tax base, but receives revenues from member local governments that are required by state law to make those contributions. There is no regional effort to direct development into Atlanta; rather, the opposite is true. Uncoordinated local land use and facility planning and weak regional planning facilitate continued outward expansion of the metropolitan area. Infill and redevelopment are seen as more problematic than simply developing further out. On the other hand, recent court cases have made it easier for developers to assemble and redevelop underdeveloped parts of the urban area, albeit without significant public assistance.

Portland is served by the Metropolitan Service District (MSD), which is a nine-member planning and administrative body that is elected from nine districts representing the urbanized areas of the Oregon side of the Columbia River. Together with strong state land use planning laws, the

MSD administers a regional urban growth boundary, works with local governments to attain regional balance in housing distribution, coordinates transportation investments, and manages particular functions such as the regional zoo, regional parks, and regional solid waste facilities. Because of MSD and state planning policy, urban development has been contained within an urban growth boundary. Infill and redevelopment are encouraged to a degree different from that in other metropolitan areas simply because a discrete boundary has been placed around urban development.

DIFFERENCES IN CENTRAL-CITY DEVELOPMENT

Table 1.5 illustrates major population and employment trends among the Atlanta and Portland central cities in the context of metropolitan growth between 1960 and 1990. Table 1.6 reports trends in densities, families, and housing costs. Table 1.7 shows trends in local government revenue, local government expenditures per capita, and capital debt.

On balance, the Portland central city has enjoyed more growth and development relative to its metropolitan region than has Atlanta. Portland has gained population, whereas Atlanta has lost population. Portland has added almost twice as many housing units to its stock as has Atlanta, 55,327 to 28,619. Even more dramatically, Portland has gained more occupied housing units than Atlanta, 52,412 versus 9,799. Between 1967 and 1987, Portland lost fewer manufacturing jobs than Atlanta, both in numbers and proportion: 8,300 jobs or a 20.6% loss versus 13,700 jobs or a 28.7% loss. During the same period, Portland gained greater regional share in retail sales than Atlanta, 24.1% versus 10.2%. It also had an increase in 4,534 retail employees during that period, whereas Atlanta lost 742. Although both cities added about the same total employment during 1963-1987, the Portland central city held on to a higher share of employment than did Atlanta.

Both Portland and Atlanta make substantial investments in economic development and services. However, Atlanta's per capita government expenditures exceed those of Portland: In 1987, the respective figures were $1,489 and $897. Atlanta also has vastly more public debt than Portland: In 1987, Atlanta owed $1.23 billion and Portland only $550 million. Although Atlanta seems to be making greater efforts to invest in and stimulate revitalization, Portland has been more effective. Why? The ingredient missing from Atlanta's revitalization efforts appears to

TABLE 1.5 Central-City Growth, Atlanta and Portland Central Cities, 1960-1990

Characteristic	1960	1990	% of Metropolitan Area 1960	% of Metropolitan Area 1990
Population				
Atlanta	487,455	394,017	39.06	13.91
Portland	372,676	437,319	43.62	29.59
Black population				
Atlanta	186,464	264,262	64.42	35.90
Portland	15,652	33,530	96.10	80.80
Manufacturing employment				
Atlanta	52,441	40,300	42.38	20.11
Portland	35,656	32,700	51.45	29.49
Wholesale employment				
Atlanta	33,074	25,328	76.27	21.42
Portland	19,757	24,284	77.37	51.98
Retail employment				
Atlanta	40,753	46,713	62.18	18.76
Portland	28,398	36,681	58.58	31.64
Service employment				
Atlanta	23,477	90,984	74.77	34.01
Portland	13,352	47,381	70.43	46.57
Total employment				
Atlanta	237,977	300,448	56.71	24.35
Portland	146,385	202,244	62.92	38.40

be regional development management that directs development opportunity toward the center rather than facilitating its continued outward expansion.

Atlanta Revitalization

Atlanta's urban revitalization efforts are, for the most part, ineffective relative to the costs involved. The problem is lack of regional consensus on the role of Atlanta in accommodating regional development. In this section, we will review the major revitalization efforts and assess, where possible, the effectiveness of those efforts.

TABLE 1.6 Trends in Densities, Families, and Housing Costs, Atlanta and Portland Central Cities, 1960-1990

Characteristic	Year	Year	% Change
Density per square mile	1960	1990	1960-90
Atlanta	3,584.2	2,989.5	-16.59
Portland	4,192.1	3,839.5	-8.41
Per capita income ($)	1974	1987	1974-87
Atlanta	4,527	11,689	158.21
Portland	5,192	11,830	127.85
Families	1970	1990	1970-90
Atlanta	119,326	86,737	-27.31
Portland	96,921	103,967	7.27
Families below poverty	1969	1989	1969-89
Atlanta	18,900	21,200	12.17
Portland	7,800	10,100	29.49
Total occupied housing units	1960	1990	1960-90
Atlanta	145,953	155,752	6.71
Portland	134,856	187,268	38.87
Median value, owner-occupied units ($)	1960	1990	1960-90
Atlanta	12,000	71,200	493.33
Portland	10,800	59,200	448.15
Median gross rent per unit ($)	1960	1990	1960-90
Atlanta	65	342	426.15
Portland	71	340	378.87

HISTORICAL SUMMARY

Atlanta has experienced three distinct phases of economic development and revitalization since World War II (Stone, 1989). The first of these eras peaked in the middle to late 1960s under Mayor Ivan Allen, whose political regime emphasized massive redevelopment and urban renewal throughout the city's economically depressed neighborhoods, called "High Priority Development Areas" (Stone, 1989, p. 60). This included the clearance of undesirable land plots for large-scale project renewal to make way for, among other things, the Atlanta Fulton County Stadium and the Atlanta Civic Center.

TABLE 1.7 Trends in Local Government Finances, Atlanta and Portland
Central Cities, 1962-1987

Characteristic	Year	Year	% Change
Local government revenue ($)	1962	1987	1962-87
Atlanta	52,086,000	641,702,000	1,032.00
Portland	41,912,000	342,995,000	618.37
Local government expenditures/capita ($)	1962	1987	1962-87
Atlanta	133	1,489	919.55
Portland	102	897	779.41
Local government debt ($ thousands)	1962	1987	1962-87
Atlanta	152,192	1,234,237	710.97
Portland	43,266	554,548	1,181.72
Local government debt/ capita ($)	1962	1987	1962-87
Atlanta	312	2,532	711.54
Portland	116	1,488	1182.78

Many renewal efforts failed during this period, leaving some city
plots vacant or underutilized. The majority of vacancies surrounded
downtown Atlanta. It would be a decade before these plots would be
dealt with through urban policy solutions such as enterprise zones. The
Metropolitan Atlanta Rapid Transit Authority (MARTA), highway and
airport improvements, and the Omni Sports Arena took shape in the
early 1970s. Although central-city revitalization was studied and Community Development Block Grant (CDBG) and Urban Development
Action Grant (UDAG) money was allocated, those efforts took a more
conservative path. Private developers and city officials were weary of
proposals and projects that attempted to revitalize the city's distressed
areas. Instead, development efforts were consolidated at the highway
transportation nodes at Atlanta's perimeter and beyond.

As development continued beyond the city's border, nonprofit organizations such as Central Atlanta Progress (CAP) and the Atlanta Economic Development Corporation (AEDC) and public agencies such as
the Atlanta Housing Authority (AHA) studied the revitalization needs
of a neglected downtown. The landmark Peachtree Plaza is an example

of a joint effort between the city and members of the business community to bring viable economic development back to the central city (George Aldridge Jr., associate director, Research, Mapping and Administration Division, Atlanta Bureau of Planning, interview, October 15, 1992).

The current era is marked by investment in a festival marketplace, the formation of enterprise zones, and important investments by the state in tourist development.

UNDERGROUND

Perhaps the most-studied and best-documented revitalization project in Atlanta is Underground, located in south-central downtown Atlanta. This specialty retail center or "festival marketplace" opened in 1989 (Sawicki, 1989, p. 349). The site is of great historical significance to the city and lies at the geographic crossroads of downtown. The first revitalization effort in this area came in 1961, when local entrepreneurs redeveloped the former railroad gulch into a retail shopping district. However, by 1981 this first effort had failed, leaving behind underutilized retail space and idle land. This set the stage for the larger development to follow.

Numerous actors, both public and private, played crucial parts in the planning and development of the new Underground. Most sources agree that Mayor Andrew Young was largely responsible for the initial political effort to redevelop Underground (Stanley, 1992). He hoped that a new and improved retail center would help ease the rising crime and urban blight in the region while also creating a much-needed social and economic center. Successful festival marketplaces were already thriving in such cities as Boston, Baltimore, and New York. These success stories were largely the products of the Rouse Company. In 1982, Rouse-Atlanta joined the city in a joint effort to redevelop Underground.

Other key private and public players involved in the project include the Downtown Development Authority (DDA), the nonprofit Central Atlanta Progress, the private Underground Festival Inc. (UFI), and the nonprofit Underground Festival Development Corporation (UFDC). The last two companies were formed for the Underground project and immediately entered into a joint contract with Rouse-Atlanta, H. J. Russell & Company, and Kingsley Enterprises Inc. For its part, UFI stated that its "purpose/reasons for development" were to "revitalize the central area; support the hospitality industry; and promote minority business development and job creation" (Stanley, 1992, p. 24).

Atlanta took responsibility for acquiring much of the land needed for Underground Atlanta as well as for financing and managing much of the project. Out of the 46 total parcels targeted for redevelopment, 41 were acquired by the city, some of which were owned by the state of Georgia (Stanley, 1992). The city also modified zoning to allow for the planned entertainment complex. Once acquired and rezoned commercial, this site was conveyed to the DDA for management and development.

Although the land was still owned by DDA, UFI purchased the commercial facilities in the developed marketplace. UFI was responsible for leasing space to Underground tenants as well as overall managerial responsibility of the facility, once completed. In 1986, Rouse-Atlanta contracted with UFDC and four other subcontractors for the physical development of Underground.

The cost of Underground Atlanta was a little more than $141 million, more than $112 million of which was funded publicly. As early as 1984, an Urban Development Action Grant of $10 million was awarded to the city for the Underground project. The most significant public funding source, however, was an $85 million bond issued by the DDA for the development of Underground and its necessary infrastructure. The DDA was given authority to issue this money through revenue bonds without needing public consent or a specific referendum. The city also issued $21 million in sales tax revenues and CDBG funds for the project. Subsequently, $6 million came from the Fulton County Building Authority for the development of parking facilities. The remaining funds were acquired privately, through the sale of UFI stock to Atlanta businesses (Stanley, 1992).

The "new" Underground Atlanta opened on June 15, 1989. It is composed of 12 acres of land, with more than 240,000 square feet of newly created leasable space. At its opening, the project included nearly 100 specialty retailers, 20 food court vendors, 22 restaurants and night-clubs, and 30 pushcarts. There were also two public plazas and 1,250 parking spaces in two parking garages. At the time of its opening, it was estimated that more than 13 million people would visit Underground in its first year. These figures put Underground in the ranks of other major festival marketplaces, such as Faneuil Hall in Boston and South Street Seaport in New York City.

Both public and private players see Underground as a vehicle for sparking economic activity and growth in a struggling downtown Atlanta. To what extent have their goals been met, and with what consequences? The most tangible effect of Underground has been its direct economic impact. At its opening, Rouse-Atlanta predicted that Underground would generate nearly 3,000 jobs; more than $5 million annually in new sales and property taxes; approximately 150 businesses, many of which are minority owned; and increased MARTA ridership (Atlanta Bureau of Planning, 1990). However, employment has been consistently in the range of 1,500 to 1,800, including summer peak hiring. This number clearly falls short of projected employment, even though it includes smaller developments not officially part of Underground (Sjoquist & Williams, 1992).

Until the mid-1990s, Underground successfully maintained the viability of its tenants. By 1992, less than 10% of original Underground businesses had folded. This trend changed, however, as a number of merchants have closed their doors in the past few years (Sjoquist & Williams, 1992). Turnover has become high. Nearly one-fifth of the Underground space is leased to short-term tenants, and vacancies have run to 9%. Moreover, Underground has been losing about $1 million a year in uncollectible rent; it lost $28 million during its first four years of operation (*Atlanta Journal-Constitution*, June 12, 1994, p. D6).

Sales at Underground were $70 million in 1989, surpassing early projections by 15%, but sales have been at or below $65 million since then. Some 43% of Underground visitors have been either tourists or convention delegates; the remaining have been either local residents or downtown workers. Thus almost half of potential sales have been generated by visitors from outside the Atlanta region.

It is difficult to assess the indirect effects Underground has had on downtown economic development. Mayor Young, Rouse-Atlanta, and others insisted that the project would act as a catalyst and spark other economic activity. One study that examined this phenomenon by looking at building permit activity and office market statistics before and after Underground concluded that "private investment is unsubstantial, indicating that the downtown market is still perceived as somewhat risky thus constraining follow-on development to Underground" (*Atlanta Journal-Constitution*, June 12, 1994, p. D7).

Although the greater economic effects of Underground are questionable, it is undeniable that the project has given downtown Atlanta a social center. Underground has become a focal point for downtown, providing a node for New Year's Eve celebrations and other community events. Although visitor figures are down, the complex still draws city residents and tourists alike. Although the effect that Underground as an urban center has had on Atlanta is not quantifiable, it is certainly important. It could even be argued that Underground fosters civic pride. As an editorial in the *Atlanta Constitution* put it, "Underground has become the city's true center, a place to share laughter, shed tears and come together" (Teepen, 1990).[1]

However, Underground's valuable cultural role hinges on its sustained economic viability. At Underground's opening, Rouse-Atlanta predicted that the venture would stabilize financially in three years. There is much argument over whether this has been the case. A 1992 economic analysis of the Underground Atlanta project noted that the project had not, up to that point, turned a profit (Sjoquist & Williams, 1992). If the worst-case scenario of "financial failure" comes to pass, this will have a measurable negative effect on the city. If Underground fails, city taxes will have to cover the $7.5 million DDA annual bond debt for 25 years, and Atlanta will lose its "town center." Thus the final effects of Underground have yet to be revealed.

ENTERPRISE ZONES

In the early 1980s, Atlanta was faced with the problem of a dramatically shrinking tax base. This problem was the direct result of the flight of many businesses and middle- and upper-middle-income residents to the north. Atlanta had experienced this trend throughout the latter half of the twentieth century; however, the problem was compounded in the 1980s by the edge city developments at Cumberland-Galleria and Perimeter.

In direct response to this economic trend, the state at the urging of Atlanta adopted the Urban Enterprise Zone Act in 1983. Subsequently, the act was amended by the General Assembly in 1986, 1988, and 1989 (Atlanta Bureau of Planning, 1992b, p. ii). The initial act was largely prepared and revised by the Atlanta Bureau of Planning in conjunction with CAP. The two organizations worked to create tax-abated regions in certain census tracts of the city for industry, business, and residence. A census tract qualifying for enterprise zone status must have a median family income of $14,800 (Atlanta Bureau of Planning, 1992b, p. 1). This requirement raised concerns among some city officials who feared that the zones would not succeed in the more depressed areas of the city. Other members of the City Council threatened not to back the Enterprise Zone Act if their districts would not directly benefit from enterprise zones. This struggle was resolved only when specific sites and developers were brought to the bargaining table by the Bureau of Planning and CAP. Three types of enterprise zones were discussed. These were developed as the result of CAP's and the Bureau of Planning's efforts to combat industrial, commercial, and residential flight.

Industrial Enterprise Zones

During 1981, General Motors made numerous threats to relocate its Lakewood manufacturing plant. Such a move would have had a negative impact on the local economy, depleting more than 3,000 jobs and millions in tax revenue. In direct response to these threats, the Bureau of Planning proposed to create a tax-abated zone for the Lakewood plant. Ironically, even after this zone was established, Lakewood was shut down owing to the troubled national economy. Four more industrial enterprise zones (IEZs) have since been created.

Atlanta's industrial enterprise zones offer tax exemptions over a 25-year period. Abatements for real property are set at 100% for the first 5 years, decreasing 20% at the end of the following 5-year periods. Personal property taxes are abated at 100% for the full 25 years. Whereas real property, personal property, and some business inventory taxes are eligible for tax abatement, nonexempt real property is taxed by the city and county. George Aldridge, assistant director of planning for the city of Atlanta, has stated that tax-abated land "offered and continues to provide great incentives for business" (interview, October 15, 1992).

All five industrial enterprise zones were developed between 1983 and 1985 with specific approval from both the mayor and the city council. The first of these zones was approved in December 1983 as the incentive for keeping General Motors in Atlanta. The General Motors Lakewood plant remained operational until its 1990 closing. Currently, a recycling company occupies the site. With a total of 83.15 acres, the Lakewood site is the smallest of the zones under survey. It is also the least occupied.

The Atlanta Industrial Park is the largest zone in Atlanta, totaling more than 398 acres. The $60 million park's location is an important factor in its size and success. The Bureau of Planning argues that the park's proximity to substantial industrial development in Atlanta is what makes it so successful (George Aldridge, interview, October 15, 1992). Currently, most of the park is occupied.

Southside Industrial Park, the youngest industrial enterprise zone, was adopted by the city council and the mayor in May 1985. The $15 million park has 260 net acres, 60% of which is still vacant and for sale.

The AEDC has been deeply involved in developing three of these sites and, along with the Chamber of Commerce, has actively recruited businesses since their opening. Atlanta's IEZs have had mixed success over the years in generating new employment. The most successful IEZ is the Atlanta Industrial Park. In 1991, 47 companies operated in the park, and they employed 1,913 people. The other IEZs have not been as productive. The Southside Industrial Park had only seven companies in 1991, employing 398 employees, whereas the Lakewood site boasts only one firm employing a mere 22 people (Atlanta Bureau of Planning, 1992b).

Both the city and the AEDC argue that these tax-abated zones are a strong enticement for business and job opportunity in Atlanta. However, both also emphasize that many of the jobs in these zones are not "created," but rather are relocated from other sites. Our own estimate is that only 15% of the total workers are city residents. We also estimate that total property taxes to be abated in these IEZs will exceed $20 million in 1995 dollars. For the nearly 400 workers who are also city residents, the subsidy comes to about $50,000 per worker.

Housing Enterprise Zones

The adoption of the Atlanta Urban Enterprise Act in 1983 also created opportunities for housing enterprise zones (HEZs). As was true for industry, the Bureau of Planning and Central Atlanta Progress developed specific legislation for the development of these residential tax-abated properties. These two agencies worked together to locate appropriate sites and interested real estate developers. A minimum eligible HEZ plot size, set at five acres, as well as considerable city council opposition, prolonged their search. However, by October 1986 the first suitable site and developer had been named. Soon after, an ordinance was adopted and approved, and construction and development began at the first site.

Between 1986 and 1993, the city created 15 HEZs. Six of these sites are clustered in the Bedford-Pine region of Atlanta. In 1986, much of the land in this neighborhood was vacant and publicly owned, owing to numerous urban renewal efforts in the 1960s. One "successful" urban renewal project was the Atlanta Civic Center, which still exists in Bedford-Pine today. However, the slum clearance and land speculation of the 1960s has left idyllic sites for housing enterprise zones, situated in the vicinity of the civic center (George Aldridge, interview, October

15, 1992). The remainder of the zones are located to the south and west of downtown.

Atlanta's HEZs abate all nonexempt, real property taxes for a period of 10 years over the following schedule: 100% exempt for the first 5 years; 80% for years 6 and 7; 60% for year 8; 40% for year 9; and 20% for year 10 (Atlanta Bureau of Planning, 1992b). All nonexempt real property is taxed by both the city and the county. Like IEZs, sites must qualify by being located in an appropriate census tract. A site must also be a minimum of five acres to qualify as an HEZ.

Another important feature of HEZ development is land acquisition. Out of a total 15 sites, 8 were privately owned parcels prior to development. The remainder were owned either by specific neighborhood coalitions or by the city of Atlanta. This meant that real estate developers could obtain this land at less than market value. Often the city conveyed or deeded the land to the developer in question at no cost. Such was the case with the West End HEZ in 1991.

To date, the HEZs have created more than 2,890 new and rehabilitated units on more than 179 acres of land. These units range in type, size, and price, depending on the development. Except for one single-room-occupancy development, all sites are targeted for middle- to upper-middle-class residences. It is our estimate that about $20 million in property taxes have been abated in the HEZs to date, averaging about $7,000 per unit.

HEZs appear to have succeeded as an economic incentive for the development of luxury housing stock in central Atlanta. All but one of the HEZs involve apartment, condominium, or housing developments geared toward middle- to upper-middle-income families. The price range for apartments is $500 to $1,500; condominiums and single-family homes range from $60,000 to $200,000. The occupancy rates of these housing developments vary depending on the site, but none of the projects has failed. In fact, all such projects have occupancy rates of more than 80%. Some have reached maximum occupancy.

Although it is evident that HEZs have been a successful vehicle for increasing the professional residential population within downtown Atlanta, it is not yet clear what effect this will have on future economic activity. Proponents of enterprise zones strongly believe that the new residential population will create a demand for other economic activity in these otherwise "depressed" census tracts. To date, however, there are few examples of such a trend. The major site is an upscale retail mall developed in Bedford-Pine to serve the "new residential market" (George Aldridge, interview, October 15, 1992). Ironically, that mall has lost approximately 70% of its tenants since its opening. Although some argue that this retail failure is the result of poor design, unwise management, and "snooty" marketing, it certainly raises questions concerning the effects of HEZs. Although these zones have fostered gentrification, it is not yet clear whether they will help to shape these neighborhoods into economic and cultural centers.

Commercial Enterprise Zones

The city's commercial enterprise zone (CEZ) policy has the same tax abatement structure as the HEZs. Also, like the industrial zones, the CEZ has a 15-acre minimum requirement, which was established by the Georgia State Assembly. No CEZs have been created to date. George Aldridge attributes the failure in CEZs to a lack of financial support given the acre and census tract requirements mandated by enterprise zone legislation (interview, October 15, 1992). There is no indication of any CEZ development in the near future.

THE WORLD CONGRESS CENTER

The Georgia World Congress Center (WCC) plays a crucial role in Atlanta, a city known nationally as a convention hub. The WCC is the second-largest convention center in the nation, drawing millions to Atlanta each year. Since its opening in 1976, the WCC has seen considerable expansion and development, including the Georgia Dome. The Dome is not only the home of the Atlanta Falcons football team, it also provides a convention venue of unprecedented size in the city.

Although the WCC requires no operating subsidy from the city of Atlanta or the state of Georgia, it is far from a private venture. Both the city and the state have contributed to the construction and maintenance of the WCC. Consequently, the convention center has returned massive economic benefits. However, as with most large venues, considerable negative effects have been absorbed by the neighborhood that bears its weight.

The WCC was constructed in three distinct phases, each receiving extensive city or state input. The initial two phases were funded through state general obligation bonds. The first development began in 1975, at a cost of $35 million. The second development began in 1985, at a cost of $103 million (Georgia World Congress Center, 1992). Both actions required specific Georgia State General Assembly approval in order to allocate sufficient funds. In essence, most of these bonds were floated through state-sponsored programs. Recently, the state approved another expansion expected to cost approximately $75 million.

The construction of the Georgia Dome relied on specific revenue bonds with no taxpayer obligation rather than general obligation bonds. A total of $200 million for construction was dedicated to the Dome project. The revenue bond payments are met partly through city and county transit and occupancy taxes. Specifically, 2.5 cents out of every 7 cents are utilized for revenue debt service (Georgia World Congress Center, 1992). The remainder of the payments are met through various long-term leases at WCC (such as that of the Atlanta Falcons), parking revenue, and other income sources.

Most of the parcels in question during all phases of WCC construction were vacant or underutilized lots owned by the city. However, some displacement of low-income housing was necessary to provide the space needed for the complex. The Dome in particular required the acquisition of a number of privately owned parcels. This was accomplished with $14 million in revenue bonds.

The various phases of WCC development created more than 2.6 million square feet of exhibit space. The convention complex attracted more than 1.3 million visitors in fiscal year 1990-1991, involving more than 60 large trade shows and conventions as well as numerous small meetings, which surpassed the convention industry average for similar facilities. More than half of the attendance was from out-of-town visitors.

Overall, the WCC can accommodate events to 1997 at about 87% of capacity. Such a high capacity rate has helped the center to operate successfully above its costs. For example, the overall revenue of the WCC was more than $18 million in 1990-1991. The center is also an important employment provider in Atlanta. In 1992, the WCC had 269 authorized full-time positions and more than 80 part-time employees at 20 hours per week (Dan Graveline, interview, October 18, 1992).

Atlanta's reliance on the hospitality industry is clearly shown in the WCC example. Dan Graveline, director of the Georgia World Congress Center Authority, candidly stated in an interview that the WCC brings in people and money that would otherwise not be in Atlanta. It is estimated that visitors to the WCC typically spend $214 per day on food, hotel accommodations, transportation, and other miscellaneous categories. These expenditures translate into more than $580 million "new dollars" pumped into the economy, generating $39 million in state taxes alone in 1991.

The economic impact of WCC is so great that other supporting facilities, such as hotels and restaurants, are contingent on the center's success. For example, the Marriott Marquis, a large-scale luxury hotel located in downtown Atlanta, was not built until developers were assured that WCC's second phase was approved by the state (Dan Graveline, interview, October 18, 1992).

The Dome as a sports and music venue brings additional economic activity downtown. The facility can hold more than 80,000 people; it was the site of the 1994 Super Bowl and will be an important Olympic venue in 1996. The limited parking for the Dome (currently only 5,000 spaces) forces visitors to take advantage of MARTA, thus providing additional revenue to the city while creating much-needed pedestrian flow downtown.

In addition, the total employment of the WCC is approximately 54% minority, compared with the city average of only 23%. However, this is shadowed by the greater negative effects the WCC has had on minority neighborhoods. The massive complex makes its home in Vine City, a historically black community. Many residents were physically displaced by the development of the World Congress Center and the Dome. In addition, the lack of parking has resulted in immense traffic problems during Dome events.

To combat the negative effects of the WCC, specifically the Dome, a $10 million trust fund was established. This fund was earmarked for low-interest loans, housing and neighborhood revitalization, and small-scale commercial development. Specifically, $5 million was allocated for single-family homes, $3 million for multifamily residences, and $2 million for other development. Atlanta's business community was to raise 20% of the trust fund, but to date that money has not materialized. As a result of this failure to raise sufficient funds, only 42 out of a targeted 200 homes have been completed. The neighborhood is still rife with crime and blight even as millions of visitors convene nearby. Perhaps the WCC has been successful as an economic catalyst for the city at large, but it has done little to ease the economic and social woes of Vine City. Many question how long this dichotomy can continue to exist.

FUTURE REVITALIZATION
AND DEVELOPMENT IN ATLANTA

The new urban framework for economic development in Atlanta will center on the upcoming Olympic Games. The city of Atlanta's Comprehensive Development Plan estimates the economic impact of the Games at $3.48 billion over 6 years. The Atlanta Committee for the Olympic Games predicts more than 600,000 out-of-town spectators during the event, injecting more than $550 million into the city economy (Atlanta Bureau of Planning, 1992a).

The Atlanta City Council has targeted High Priority Development Areas for economic development to prepare for the international event. These six neighborhood areas will be intensely studied for economic development potential and will be eligible for UDAG and CDBG money. Already, the Olympic Stadium and other venues are in advanced stages of planning and development. Planners, developers, and city officials agree that the Olympics will provide an unprecedented opportunity in Atlanta's history. The revitalization efforts surveyed in this study will play a role in the success of the Olympics.

In the main, however, revitalization efforts of the city have been expensive, often ineffective, and in some instances phantom. Atlanta is perhaps one of the nation's leaders in revitalization efforts, but its success has been hampered by a lack of regional consensus on the city's role in accommodating regional development needs.

Portland Revitalization

In this section we will survey recent revitalization efforts undertaken by the city of Portland, Oregon. Despite considerably smaller investments in revitalization relative to Atlanta, and despite being of similar size but having much smaller regional growth from which to attract, Portland's revitalization efforts can be characterized as more effective than Atlanta's. We surmise that the difference is a regional consensus on the role of the central city in accommodating regional development.

HISTORICAL SUMMARY

Portland's efforts toward central-city revitalization have given it an international reputation among midsized cities. In a nationwide survey of urban planners, the city ranked third in development and potential and first overall in management of urban design (Langdon, 1992). Portland's commitment to revitalizing its downtown, preserving its historic districts, improving and restructuring its housing stock, and encouraging new development is evident in the aggressive planning that has occurred throughout the years. For example, the Portland City Council has adopted 16 comprehensive urban renewal plans in the past 30 years. Of those plans, 8 have been completed and implemented to date; these include urban renewal surrounding Portland State University, Emanuel Hospital, and six residential areas. As a member of the Portland Development Commission staff stated in an interview, "Portland has a successful reputation for city planning because the plans that are developed are being implemented" (December 28, 1992).

Portland's revitalization successes are attributable to high-profile government intervention in the development market. In 1958, a voter referendum created the Portland Development Commission (PDC) and charged it with using public and private resources to revitalize blighted areas of the city. The PDC is a public nonprofit corporation that enjoys relative flexibility in undertaking complex developments. Together with city planning staff, PDC staff prepare long-term urban renewal plans for targeted areas.

During Portland's early years of urban renewal—the late 1950s to early 1960s—the PDC condemned and acquired 110 acres of land adjacent to and south of downtown. An urban renewal plan was prepared for the area, and over the past 20 years it has been implemented. As a result, the South Auditorium Urban Renewal Area was redeveloped in line with city plans, and property values were dramatically increased (Bob Clay, Portland Bureau of Planning, interview, December 29, 1992). Although much of the success of this project is attributed to the fact that 97% of its funding came from private investment, it would not have been possible without the PDC and the city (Portland Development Commission, 1992b).

In 1972, the city directed its revitalization efforts to the downtown. At that time, downtown Portland faced many of the problems other cities were wrestling with. Portland's approach to downtown revitalization was an aggressive plan that proposed automobile restriction, public transit, and intense redevelopment. For example, the city recognized the value of the Willamette River to downtown and thus replaced an expressway with Governor McCall Waterfront Park in the mid-1970s. An extensive and ongoing improvement program for the park has helped it become one of the nation's best central-city waterfront areas.

It was also during this period that the design and structure of downtown Portland significantly changed. The Downtown Plan spawned many impressive architecture and urban design efforts, and created more than 30,000 new jobs and $1.5 billion in investment (Portland Development

Commission, 1992b, p. 9). Between 1974 and 1984 the downtown Transit Mall was developed. It includes dedicated bus lanes and light rail tracks (the impact of public transit is discussed in more detail below). The city's housing stock was also targeted for revitalization during the 1970s. PDC and city staff estimated that 70% of Portland's housing was more than 35 years old, and that more than 25,000 units did not meet minimum housing codes. There is now an ongoing program of rehabilitation sponsored by the city and the PDC. To date, more than 20,000 homes and apartment units have received low-interest loans for rehabilitation (Portland Development Commission, 1992b, p. 9).

Portland has enjoyed much success in combating the residential, commercial, and industrial flight that has plagued many of the nation's cities. The city's recent past presents an excellent track record in planning and implementing urban revitalization. Some of these programs continue to be carried out today. In the following subsections, we will examine Portland's central-city revitalization in more detail by looking at the inputs, outputs, and effects of specific areas, beginning with the Transit Mall and light rail system.

THE PORTLAND TRANSIT MALL
AND METROPOLITAN AREA EXPRESS

Portland's Transit Mall and light rail system, the Metropolitan Area Express (or, more simply, MAX), are crucial parts of downtown development. Buses in combination with MAX compose much of Portland's extensive public transit system in the metro area. Many argue that successful mass transit plays a vital role in central-city economic development. We examine this assertion by surveying Portland's Transit Mall development and MAX and show that the mall and the light rail system have had significant positive effects on the downtown and have helped to make conditions there conducive to revitalization.

The Transit Mall combines transit mall and shelters with restrictions on automobile use intended to give preferential treatment to buses. Pedestrian access is facilitated with widened sidewalks, comprehensive transit maps, information kiosks, and street furniture. The end result is a concentration of transit service in a specified area. In their impact study on the Transit Mall, Dueker, Luder, and Pendelton (1982) note that it creates a "special focus in the downtown area that brings people together . . . and stimulates development in a pattern that can be better served by transit" (p. 111).

Transit Mall planning began in 1970 as part of a larger plan for downtown Portland revitalization. The idea of a mall was included in other environmental and development studies and gradually evolved into an integral element in the city's downtown vision. Following a feasibility study conducted by the Tri-County Metropolitan Transportation District of Oregon (Tri-Met) in 1973, designs were solicited in 1975, with full operation achieved by 1978.

The Transit Mall consists of two exclusive bus lanes along Fifth and Sixth Avenues. Each lane stretches out for 11 blocks and hence virtually eliminates private vehicular traffic through the heart of downtown. Sidewalks were widened by roughly 10 inches, and 32 bronze-and-glass-roofed bus shelters were placed on the mall, one at each stop. Also included in construction and more recent improvements have been numerous sculptures, trees, fountains, and graphic displays.

The light rail system in Portland evolved as part of a regional strategy to balance the area's transportation system, reduce traffic congestion, help contain growth, and foster dense and more favorable development. In 1978, Tri-Met concluded that light rail technology would be the most cost-effective means of meeting these regional goals (Arrington, 1992). Cooperation between affected local governments led to the timely acceptance of the concept by the Oregon Legislature in 1979. By 1986, the "East Side" line was operational, connecting downtown to the suburbs east to Gresham, 15.1 miles away. A total of 26 vehicles serve 29 stops along the route.

The Transit Mall and MAX were financed primarily through federal assistance under the Urban Mass Transportation Act (UMTA) of 1965. Some 80% of the $15 million Transit Mall was funded from UMTA, with the balance financed by Tri-Met. UMTA and other federal agencies absorbed 83% of MAX's total $214 million cost, which included the reconstruction of one segment of Interstate 84 in the Portland area. Much of the federal money had been earmarked for a new freeway through Portland's southeast neighborhood, but when the city was successful in having state and federal highway agencies cancel the building of the freeway, the money remained available to finance the light rail development.

The Transit Mall and MAX have had profound impacts on central-city redevelopment by helping to reduce urban sprawl significantly. The 1972 Central City Plan emphasized the importance of mass transit in its downtown development strategy. The fruition of this idea in the Transit Mall has helped to maintain downtown as the dominant office employment center in the metropolitan area. However, there is some question about the direct effects that transit projects like the Transit Mall had on development. For example, a report on a study conducted by the Portland City Planning Bureau (1991) that traced development trends between 1960 and 1980 emphasizes the importance of not making a direct correlation between the Transit Mall and economic trends.

The period after mall construction saw the largest increase in new and renovated office and commercial space in the downtown area. In the 5 years after mall construction, an unprecedented 300,000 square feet of office space was renovated. Nearly a million square feet of additional new office space was added directly on the mall, and another million square feet were added within easy walking distance of the mall.

Although it may be difficult to determine the extent to which these investments would have been made anyway, there seems to be some consensus that the location of investments near the mall is not coincidence. Moreover, the mall's symbolic impact is evident in the opinion of business owners and city residents. According to one study, mall construction represented for many people a firm commitment to downtown on behalf of city government (Dueker et al., 1982). In many firms' opinions, the mall has also stimulated private investment. Whatever the case, the Transit Mall was the fulfillment of a significant aspect of the 1972 Downtown Plan, one that vigorously pursued central-city development.

MAX's impact on economic development has been far more than symbolic. Since its construction, the light rail has spawned a corridor of economic development and revitalization. Existing stores along the MAX line at the beginning of operation reported almost instantaneously higher sales and increased weekend foot traffic as a result of light rail ridership. Both Portland and Gresham stores reported increases in sales volumes ranging from 20% to 50% in MAX's first year of operation. In general, more than $800 million in development has occurred along the light rail line, and more than $400 million is planned for the near future.

Regardless of whether one accepts the contention that mass transit equals increased economic development, there is no denying that the Transit Mall and MAX have played their part in Portland's central-city revitalization. In 20 years, the number of people working downtown has increased from 59,000 to 94,000. The total amount of office space has increased from 5 million to 14 million square feet. More than 5 million square feet of retail space, supporting more than 1,000 stores with more than 10,000 employees, has been built since these transportation improvements have been in place (Arrington, 1992). Public transit is partly to thank for these figures.

URBAN RENEWAL AREA DISTRICTS

What else has Portland done to achieve its level of success? The latest Central City Plan was adopted by the Portland City Council in 1988 and was designed to shape development for the next 20 years. Included in the plan was the call for improvements for the Transit Mall and part of the light rail line. The plan also defined seven urban renewal zones as having priority for urban economic development. Two of these zones, Downtown Waterfront and Northwest Front Avenue, have been urban renewal areas since the 1970s.

The Transit Mall and MAX have had profound effects on urban renewal and central-city revitalization in Portland. They reflect strong intentions by both city and state officials to control and guide urban development through good planning and successful implementation. The areas specified in the 1988 plan were also defined to focus development efforts further in areas of Portland where such development would be desirable. Although each of these areas has specific design guidelines and economic priorities, all share common attributes.

Portland's comprehensive plan charges that all urban development shall "support regional goals, objectives and plans adopted by the Columbia Region Association of Governments and its successor, the Metropolitan Service District, to promote a regional planning framework" (Portland Bureau of Planning, 1991, p. 4). Specifically, the areas under study all support the concept of an urban growth boundary for the Portland metropolitan area. Each plan also is in concurrence with the city's established urban planning area and urban service area boundaries.

In essence, the strong regional perspective adopted by the city of Portland has become an essential part of all development conducted in the central city. Thus regional planning helps to shape urban renewal area planning, which in turn helps to determine the physical shape of the city. Such an approach implies carefully planned and well-conceived plans at every level. The urban renewal areas under study follow specific design, infrastructure, and cost guidelines as outlined by the Portland Bureau of Planning and as accepted by the city council. The similarities between these areas begin with the means by which development projects within their boundaries are financed.

The primary financing mechanism for urban development in Portland is tax increment financing (TIF), which was originally used as a means of providing local matching funds for federal urban housing projects. The Oregon State Legislature expanded the use of TIF in 1979 to include most economic development and urban renewal, thus opening the door for a much wider scope of project funding.

Simply stated, TIF is a financing vehicle through which development can be pursued that would not normally be feasible when undertaken by the private sector alone. It provides local governments the opportunity to allocate a portion of property tax revenues to promote and shape economic growth. TIF essentially freezes the assessed value of real property that is being considered for development (City Club of Portland, 1991, p. 17). Subsequently, bonds are sold to finance the public cost of development, and private development is aggressively encouraged. After development and improvement occur, increases in property taxes accrued as a result of increasing property value are used to help pay off urban renewal bonds initially used for land assembly and infrastructure development.

One crucial element of eligibility for the use of TIF as mandated by the Oregon State Legislature is that an urban renewal district and plan for that district must be established. Hence TIF can be applied only to development within that area. Central-city revitalization financed through TIF has proven to be a successful means of initiating private development in blighted areas. These areas, however, must be carefully defined and planned for, as demanded by state law. All of the urban renewal areas defined in the 1988 Central City Plan were created as qualified TIF zones. Specifically, Redevelopment Fund budgets have been created and annually updated for each of the seven zones approved by the Portland City Council. We discuss two such zones in detail below: Downtown Waterfront and the Oregon Convention Center.

Downtown Waterfront Urban Renewal District

The Downtown Waterfront urban renewal district is the largest and longest-lasting urban renewal area under study. Its development plan was adopted in 1974 and is still being implemented. Virtually all of the development that has occurred during 18 years in this area—except for the Transit Mall—has been partially financed through TIF.

TIF is used to leverage private investment. To what extent has leveraging occurred? Two of the largest projects within this area, Pioneer

Place and RiverPlace, totaled more than $240 million in private and public investment from 1987 to 1990. Overall, since 1970, more than $1.5 billion in public and private investment has been made in Portland's downtown, including this renewal area (Portland Development Commission, 1992a).

The amount of development that has occurred in this urban renewal area has been far more grand than in any of the other urban renewal areas under study. This is primarily because of the emphasis of regional and municipal plans on a vibrant and successful downtown environment. This emphasis has taken many forms. For example, a formal directive was given to the PDC from Portland's City Council in 1974 to make the central city more attractive for commercial, residential, and other uses. The result has been an onslaught of new development in Portland since the inception of the downtown urban renewal area in 1974. One of the more significant developments has been Pioneer Place, which opened in 1990. This festival marketplace was developed by the Rouse Company at a cost of about $180 million. It includes Saks Fifth Avenue as an anchor store, a 17-story office tower, and more than 80 retail and specialty shops. The development is a vital component of the revitalized Pioneer Courthouse Square, which opened in 1984. The two provide a rich blend of historic preservation and upscale shopping and dining in the heart of downtown Portland.

Other key outputs produced within this urban renewal boundary include the South Waterfront Project, which covers 73 acres of housing, office, marina, and riverfront park developments at a current cost of $66 million. Numerous housing rehabilitation and building restoration projects have been completed, including the Union Station Project, 482 single-room-occupancy and low-income housing improvements, 30 facade renovations, and parking garages supporting downtown traffic flow.

The development completed under the Downtown Waterfront urban renewal plan has helped to redefine Portland as the regional center for business, retail, government, and entertainment, as well as improved the area's housing stock dramatically. As the PDC emphasizes, design guidelines of all projects pursued in this area have helped to turn the central city into an "area favoring pedestrian and human activities" (Portland Bureau of Planning, 1990). For example, in an article in the *Atlantic*, Philip Langdon (1992) noted that Pioneer Courthouse and Place "is a center carved out like an amphitheater, with terraces of brick seating that make it a favorite site for rallies, music, and other outdoor events" (p. 136).[2]

The numerous accolades for the design and pedestrian nature of downtown are accompanied by the positive economic impact the completed projects have had in the urban renewal area. Overall, the number of jobs in the Portland area has increased by 100,000 since 1983. More than 30,000 of these jobs have been created as a result of the significant developments in downtown Portland. Undoubtedly, the role of downtown in boosting economic development is inextricably linked with the aggressive public transit campaign carried out by Tri-Met and the city at large. These two forces—development within the Downtown Waterfront urban renewal area and the supporting transit networks of the Transit Mall and MAX—have had a dramatic effect on the downtown Portland economy.

Oregon Convention
Center Urban Renewal District

If downtown Portland has been successfully transformed through central-city revitalization, the Oregon Convention Center urban renewal district (more simply called the Lloyd district) is certainly well on its way. The area was once classified as underdeveloped in terms of retail and commercial space (White, 1989, p. 6). However, the area is an important location as a gateway to downtown and as the eastern anchor of the central city. Also, MAX began its operations through this area in 1988, making it ideal for urban redevelopment. In recognition of these attributes, the Portland Bureau of Planning prepared and the city council adopted a specific plan for the Lloyd district in 1989.

The plan calls for the adoption of detailed urban design guidelines to help preserve the unique nature of the district. It also establishes priorities for economic development and methods for stimulating private sector involvement. As was true with the Downtown Waterfront district, the primary mechanism for sparking redevelopment is tax increment financing. The anchor development within this district is the Oregon Convention Center, a 17-acre facility built in 1990. It is sandwiched between the Memorial Coliseum and Interstate 5. A MAX station provides additional transportation options to employees and visitors. Its total project cost was $85 million, of which $65 million was raised from the sale of a general obligation bond and $15 million was paid by the state of Oregon. The remaining sum was paid by a local improvement district that includes most of the businesses that will benefit from the convention center.

Although the Lloyd district was created only in 1989, the area has had considerable amounts of development and there are even bigger plans for the future. The convention center houses 500,000 square feet of exhibit and meeting space. Its award-winning design and many modern amenities make it a desirable West Coast location for trade shows and conventions. The Lloyd district will also soon house the Headquarters Hotel in order to help maximize the regional potential of the convention center. A portion of this site has already been completed. The Portland Trailblazers basketball team has built a 19,000-seat arena in the district as well.

Smaller development and revitalization have taken place in the area as well under the name of Pacific Development. Specifically, the company has developed 40 out of an eligible 75 blocks in the district, with retail shops, restaurants, and other street-level establishments.

The Oregon Convention Center has had a strong impact since its opening in 1990 (Michael Smith, Portland Oregon Visitors Association, interview, January 15, 1993). Specifically, the facility attracted more than 105,850 people to Portland between September 1990 and December 1991. These visitors had an obvious impact on the economy, generating more than $39 million in service and tourist revenues. The following year was estimated to have slightly less economic impact on Portland, but the 1993 convention center schedule was predicted to inject more than $55 million into the local economy (Portland Oregon Visitors Association, 1993).

One specific impact of the convention center is the economic stimulation generated from hotel lodging. The increased demand for hotel space has led to the ongoing construction of the Headquarters Hotel in the district. It is estimated that this hotel will generate up to 2,400 jobs and as much as $110 million in annual economic benefit. The city is financing many of these improvements through TIF proceeds to ensure timely completion.

Another crucial effect of the development of this urban renewal zone is that it connects downtown to the Lloyd district across the Willamette River. This has been achieved physically, through transit and design, as well as economically, through the developments discussed above. In essence, this area is being gradually redeveloped into an east-side extension of the already successful mixed-use Downtown Waterfront district. Developers are optimistic that current facilities and planned development will not only help make the Lloyd district an extension of downtown and part of the central city, but also make the area "some of the best of what the city of Portland has to offer" (White, 1989, p. 6).

FUTURE REVITALIZATION
AND DEVELOPMENT IN PORTLAND

Portland's Central City Plan, adopted in 1988, will lead development in the region for the next 20 years. The PDC has outlined admittedly grand schemes for unifying the east and west of the city, enhancing existing urban renewal districts, and focusing attention on other regions not yet highlighted. The Lloyd district plans emphasize continued private sector investment and neighborhood development as well as plans for a new sports complex linked to the existing convention center.

Tri-Met is aggressively pursuing plans for a west-side light railway that it argues will have an overall impact of $1.9 billion on Oregon's economy. The agency has secured a 75% federal financial commitment that translates to a $680 million capital infusion for the state and the city. A north-south line running 20 miles, connecting Vancouver, Washington, and Oregon City, Oregon, is under study.

The financial future of other development and central-city revitalization is not as clear as the impending west-side light railway. Recent legislation has placed strict limitations on tax rates and has dramatically affected the use of TIF. In fact, the longevity of the PDC is also in question. This unique organization, which for years acted as the implementing organization for all of Portland's detailed Central City Plans, may be in jeopardy. Thus Portland's future success in urban renewal and development is in some doubt, given the possible loss of TIF and the PDC.

On balance, Portland seems to have been far more effective in stimulating revitalization than has Atlanta. In the next section we review the reasons for this in the context of regional urban development policy.

Policy Implications

To be sure, there are many differences between Atlanta and Portland. For example, during the period 1960 to 1990, Atlanta's regional population grew by 127%, whereas Portland's grew by 73%. Atlanta has risen to world prominence in trade, transportation, and arguably culture (with its selection as the site of the 1996 summer Olympic Games); Portland has become a mecca for urban design, architecture, and planning enthusiasts. Atlanta has 10 times the African American and minority populations as Portland, although 90% of the Portland region's African Americans and minorities live in the central city. Whereas Portland accounts for 50% more of the share of its region's employment, Atlanta has 50% more employment than Portland in the central city.

But there are interesting similarities between the two cities. They have comparable populations, land areas, and terrains. Arguably, they have similar histories as well. Atlanta began as a transportation center, was burned down during the Civil War, and was reborn during the late 1800s. Portland began as a shipping port, suffered a major fire just after the Civil War, and was also reborn in the late 1800s. The two cities' populations, economic bases, and employment levels were reasonably similar through the years ending in 1980.

Especially since the 1980s, however, Portland and Atlanta and their regions have grown differently. Atlanta has become the quintessentially sprawled U.S. city. Its suburbs have exploded. Even the suburbs have suburbs: the Atlanta Regional Commission's 1985 projection for 1990 for its seven-county region was too high by 300,000 because the outer counties grew by 300,000 more than projected by the state of Georgia (Atlanta Regional Commission, 1985). Edge cities have risen. The three largest edge cities each contain more office and retail square feet than downtown Atlanta. Most of the region's class A office space is not located in downtown Atlanta; indeed, downtown accounts for barely 10% of total office workers. It can easily be argued that state and local suburban government policies affecting the Atlanta area are clearly aimed at depopulating the city.

Portland has become the prototypical compact metropolitan region of the future. Not hemmed in by mountains and bodies of water as are other compact cities, such as San Francisco, New Orleans, and Manhattan Island, its emerging urban form is consciously pursued. Together with one of the nation's most ambitious state-level land use planning programs, the Portland metropolitan area has pursued a policy of limiting the territory over which development will occur (Leonard, 1983). But instead of imposing growth controls that limit growth rates, the Portland area has engaged in proactive growth accommodation that includes fairly aggressive redevelopment, greater emphasis on transportation system-land use linkages, higher density requirements on new construction, streamlined permitting to reduce uncertainty, and regionalized planning and management of major infrastructure such as water, sewer, and transportation (Nelson, 1991). State and regional development policy is explicitly geared to directing development into the central city (Knaap & Nelson, 1992). Indeed, downtown Portland contains the largest amount of office and retail space in the region, and accounts for more than a quarter of the region's office space (Urban Land Institute, 1992).

Two more different approaches to regional development policy may not be found. But how do those approaches affect central cities? A growing literature demonstrates that unmanaged regional development results in inefficient dispersion of development away from the central cities. Much of the inefficiency is caused by a panoply of federal and state policies that favor

1. new construction over the rehabilitation and reuse of existing buildings;
2. highway transportation over public transit;
3. the conversion of undeveloped land for urban uses over the reuse of developed urban land;
4. the construction of single-family, owner-occupied housing over multi-family and rental housing;
5. growing areas over depressed areas; and
6. new locations, recently developed, over old locations (Bourne, 1980).

The result, according to some analysts, is "urban sprawl" (Clawson, 1962), most often characterized as discontinuous, low-density development (Harvey & Clark, 1965). As Richard Peiser (1989) states: "The impact of discontinuous development on density is important because uniformly low-density urban development is inefficient. It increases transportation costs, consumes excessive amounts of land, and adds to the cost of providing and operating public utilities and services" (p. 193). Further, there are "other consequences associated with urban sprawl such as inefficient resource allocation for public facilities, increased transportation costs, and removal of agricultural land" (p. 202).

This is the development pattern that characterizes the Atlanta metropolitan area, and most other areas in the United States. The question is: What could be gained through a more compact development pattern that directs new development into central cities? The U.S. Department of Housing and Urban Development examined 106 U.S. metropolitan areas using data primarily from the early 1970s, and found the following:

1. For large and medium-size metropolitan areas with dominant CBDs, neighborhood density and diversity are associated with lower vehicle miles traveled.

2. Urban neighborhoods composed of small to medium-size multifamily units tend to have lower energy costs.

3. Low-density, dispersed patterns result in lower air pollution, although this finding is based on data collected during the middle 1970s prior to full implementation of automobile fuel economy and emissions standards—in fact, the federal Environmental Protection Agency has declared Atlanta a "nonattainment" area principally because of its automobile-dependent land use pattern, yet Portland has been removed from nonattainment status.

4. Higher-density development is associated with higher operating costs, although those higher costs are usually associated with delivery of more services that benefit more people than lower-density development patterns can economically deliver.

5. Higher density is associated with lower water consumption per capita.

6. Solid waste disposal sites are more easily accomplished with compact development patterns.

7. Farmland preservation is best facilitated with compact development patterns (Office of Policy Development and Research, 1980; see also U.S. House of Representatives, 1980).

A particular advantage of more compact development patterns is conservation of energy, chiefly fossil fuels. Newman and Kenworthy (1989) analyzed gasoline consumption among world cities and found that consumption in U.S. cities is twice the world average after differences in gasoline price, income, and vehicle efficiency are accounted for. The reason for excess gasoline consumption in the United States— and associated air pollution, socioeconomic segregation, and development of rural lands—is the predominant urban pattern. Newman and Kenworthy argue for increasing urban density, strengthening the city center, extending the proportion of cities that have inner-area land uses, providing good transit options, and restraining the provision of automobile infrastructure. They claim that these efforts can be achieved through reurbanization and reorientation of transportation priorities.

The bottom line is that more compact patterns of urban development, anchored by relatively high-density central cities, result in greater economic development. The emerging literature is beginning to show in a reasonably conclusive manner that gross domestic product is improved by both larger city size and higher densities (Banta, 1989; Cropper, 1987).

In the absence of regional or state-level coordination of development patterns that reverse the effect of sprawl-inducing policies on regional urban form, central cities are at a distinct disadvantage in attracting new investment. Atlanta and Portland offer contrasting case studies on this point. Despite far more growth in the Atlanta region, Atlanta has fallen behind Portland in terms of retaining regional prominence in housing, employment, and economic activity. To compete with urban sprawl, Atlanta has incurred considerable debt to finance activities aimed at stimulating development. With far less total debt and debt per capita, Portland has achieved greater economic returns if for no other reason than development is constrained by a regional urban growth boundary that contains urban sprawl. In the absence of regional coordination of transportation and land use, more, not fewer, roads are being built in Atlanta, and MARTA is losing ridership share because it is increasingly too far removed from new development. In contrast, Portland's fledgling light rail system is growing in both ridership and lines, principally because of regionally coordinated land use and transit planning aided by regional urban containment strategies.

In conclusion, central-city revitalization probably depends more on regional management of development than on specific investment or redevelopment activities. In the absence of regional development management, central cities will continue to fare poorly in the regional competition for new investment. Suburban and exurban communities will continue to lure new investment through a variety of inefficient federal, state, and local inducements. The result is inefficient development patterns that can only undermine long-term economic vitality for urban regions and the nation. New governance structures are needed to reverse inefficient development trends. Portland, Oregon, offers one model. What may be needed is a federal initiative to elevate the management of urban development patterns from local, usually competing, interests to the region and, in some instances, the state.

Notes

1. Reprinted with permission of *The Atlanta Journal* and *The Atlanta Constitution*.
2. Reprinted with permission of the *Atlantic*.

References

Arrington, G. B. Jr. (1992). *Portland's light rail: A shared vision for transportation and land use.* Portland: Tri-Met.

Atlanta Bureau of Planning. (1990). *Questionnaire on the Project Underground Atlanta and the Developer Downtown Development Authority.* Atlanta: Department of Planning and Development.

Atlanta Bureau of Planning. (1992a). *City of Atlanta comprehensive development plan.* Atlanta: Department of Planning and Development.

Atlanta Bureau of Planning. (1992b). *1991 annual report on urban enterprise zones.* Atlanta: Department of Planning and Development.

Atlanta Regional Commission. (1985). *Atlanta regional projections 1985-2005.* Atlanta: Author.

Banta, S. M. (1989, December). Consumer expenditures in different-size cities. *Monthly Labor Review,* pp. 44-47.

Bourne, L. S. (1980). Alternative perspectives on urban decline and population deconcentration. *Urban Geography, 1*(1), 39-52.

City Club of Portland. (1991). Report on tax increment financing in Oregon. *City Club of Portland Bulletin, 72*(2).

Clawson, M. (1962). Urban sprawl and speculation in suburban land. *Land Economics, 38,* 99-111.

Cropper, M. L. (1987). The value of urban amenities. *Journal of Regional Science, 21,* 359-374.

Dueker, K., Luder, P., & Pendelton, P. (1982). *The Portland Mall Impact Study.* Washington, DC: Urban Mass Transit Administration.

Georgia World Congress Center. (1992). *1991 annual report.* Atlanta: Author.

Harvey, R. O., & Clark, W. A. V. (1965). The nature and economics of urban sprawl. *Land Economics, 41,* 1-9.

Knaap, G. J., & Nelson, A. C. (1992). *The regulated landscape.* Cambridge, MA: Lincoln Institute of Land Policy.

Langdon, P. (1992, November). How Portland does it: A city that protects its thriving civil core. *Atlantic.*

Leonard, J. H. (1983). *Managing Oregon's growth.* Washington, DC: Conservation Foundation.

Nelson, A. C. (1991). Blazing new planning trails in Oregon. *Urban Land, 49*(8), 32-35.

Newman, P. W. G., & Kenworthy, J. R. (1989). Gasoline consumption and cities: A comparison of U.S. cities with a global survey. *Journal of the American Planning Association, 55*(1), 24-37.

Office of Policy Development and Research. (1980). *Metropolitan development patterns: What difference do they make?* Washington, DC: U.S. Department of Housing and Urban Development.

Peiser, R. B. (1989). Density and urban sprawl. *Land Economics, 65,* 193-204.

Portland Bureau of Planning. (1990). *Central City Plan: Fundamental design guidelines.* Portland, OR: Author.

Portland Bureau of Planning. (1991). *Comprehensive plan: Goals and policies.* Portland, OR: Author.

Portland Development Commission. (1992a). *Annual urban renewal report.* Portland, OR: Author.

Portland Development Commission. (1992b). *Report: Moving into the fourth decade.* Portland, OR: Author.

Portland Oregon Visitors Association. (1993). *Portland/Oregon Visitors Association convention and trade shows confirmed bookings committed to OCC as of February 8, 1993.* Portland, OR: Author.

Sawicki, D. S. (1989). The festival marketplace as public policy: Guidelines for future policy decisions. *Journal of the American Planning Association, 55,* 347-361.

Sjoquist, D. C., & Williams, L. (1992). *The Underground Atlanta project: An economic analysis.* Atlanta: Georgia State University, College of Business Administration.

Stanley, L. (1992, June). *An assessment of the development impacts of Underground Atlanta.* Paper prepared for City Planning Program, Georgia Institute of Technology, Atlanta.

Stone, C. (1989). *Regime politics: Governing Atlanta, 1946-1988.* Lawrence: University Press of Kansas.

Teepen, T. (1990, September 20). No doubt now about Underground's value. *Atlanta Constitution,* p. A18.

U.S. Bureau of the Census. (1991). *1990 census of population and housing* [database]. Washington, DC: Author.

U.S. House of Representatives, Committee on Banking and Finance. (1980). *Energy and compact urban form.* Washington, DC: Government Printing Office.

Urban Land Institute. (1992). *Market profiles 1992.* Washington, DC: Author.

White, M. (1989, December 11). Lloyd district growth will totally change city. *Downtowner.*

2

Baltimore and the Human Investment Challenge

DAVID IMBROSCIO
MARION ORR
TIMOTHY ROSS
CLARENCE STONE

In this chapter, we describe the human investment challenge that Baltimore faces and consider how changes in federal policy could enhance the city's capacity to meet this challenge. Baltimore is perhaps best known for its long and successful efforts in physical redevelopment, especially in the Inner Harbor area. Less well known are the city's efforts at human renewal and the constraints on those efforts. It is particularly important that we spotlight human investment, because the Baltimore experience demonstrates that physical redevelopment by itself is not enough to revitalize an urban community. Human investment is a necessary element in any overall strategy of renewal.

Human Investment Defined

In basic terms, human investment is an effort to make people more productive members of society. It is centrally concerned with the inculcation of skills and work-related aptitudes; thus education and job training are its mainstays. Human investment also includes measures to promote health, and it encompasses day care and preschool programs as well.

An aim of human investment is to promote economic self-sufficiency in a technologically complex world. Such a policy does not foreclose a social responsibility to provide for those who are temporarily or even permanently dependent, but human investment itself is guided by the desirability of making economic self-sufficiency as broad based as possible. It entails the use of public resources to provide through programs what, for many children, is not provided by family and neighborhood. Put another way, the path to economic self-sufficiency is more difficult and more strewn with obstacles for some than for others. Human investment involves efforts to remove such obstacles.

Central cities face a particularly strong challenge in this regard. The economic and demographic changes they are experiencing constitute numerous obstacles to economic self-sufficiency for their citizens and at the same time limit the resources available to these cities. Baltimore is an appropriate case study. Both the problems the city faces and the resources available for the city to draw on are shaped by fundamental economic and demographic changes.

A Brief History

Baltimore is an older industrial city struggling with the transition to a predominantly service economy. In this process it has lost population and business activity as its suburbs have grown and prospered. Paralleling trends at the national level, manufacturing jobs in Baltimore have declined sharply since World War II. On the other side of the ledger, service jobs have replaced manufacturing jobs in quantity—but not in quality. The city's growth in service jobs was especially sharp after 1970, as Baltimore's Inner Harbor became the site for new offices and hotels.

With the shift from an industrial to a service economy, net job growth in Baltimore has been quite modest, and the quality of entry level-jobs, in particular, has declined. Even more vigorous growth is no guarantee that city residents will enjoy high levels of employment. At present, half the jobs in the city are held by suburban commuters. Only investment in education and job training for city residents combined with an effective antidiscrimination policy will open up jobs to all segments of the city population. The following brief history puts Baltimore's experience into context and shows how the current situation came into being.

Baltimore has been the center of urban life in Maryland for much of the state's history. Founded in 1729, the city began as a center of commerce with wheat and flour trading. In the mid-1800s, the Industrial Revolution pushed Baltimore's economy from commerce and trade to one with an emphasis on manufacturing. Nineteenth-century residents of Baltimore's blue-collar communities could find work in foundries, machine shops, and factories. Throughout the early years of the twentieth century Baltimore continued to add to and diversify its industrial base. The city's port facilities were also a large part of the local economy, providing jobs for hundreds of workers.

The growth of industry attracted people to Baltimore in search of job opportunities. During the nineteenth century, European immigrants arrived from Ireland, Germany, and later Italy and Poland, along with smaller numbers from Russia and Lithuania, making Baltimore a city of ethnic neighborhoods. Today, however, immigration from abroad has ceased to be a major influence in Baltimore's demography; the "new immigrant" groups from Asia and Latin America make up less than 3% of the city's total population. Baltimore's current population is composed primarily of whites and African Americans.

Historically, slavery was never an integral part of the city's economy, and at the outbreak of the Civil War Baltimore had the nation's largest free African American population. Still, when the Civil War came to a close, African Americans were a small minority, only 15% of the city's population.

Despite the heritage of a free African American population, post-Civil War race relations in Baltimore were not radically different from those of cities in the South. Exclusion from many occupations, efforts at disenfranchisement, and Jim Crow practices put Baltimore's small African American population at a severe disadvantage. Nevertheless, within the confines of segregated life, Baltimore's African American community developed a significant and assertive middle class. Civil rights activism began in the 1930s, even though Baltimore's African American population was still less than one-fifth of the city's population at that time.

World War II and the search for wartime jobs brought the African American proportion of the population beyond the 20% mark. African American growth was, however, restricted to the central city. Housing discrimination delayed African American movement to the suburbs until after a pattern of racial concentration in the central city was already established. Even when chances in the central city for stable employment at decent wages diminished, African Americans had little choice but to cluster in cities such as Baltimore, where they already had a residential base.

As good-paying jobs for blue-collar workers decreased, a human investment strategy of increased education and targeted training would have made enormous sense. Instead, national urban policy, from the 1949 Housing Act on, concentrated on the physical redevelopment of the city.

Baltimore was an early participant in urban renewal, and after the election of Mayor William Donald Schaefer in 1971, the city accelerated its massive project of downtown revitalization primarily through redeveloping the Inner Harbor area. By the 1980s, Baltimore gained prominence as a "renaissance city." The city reshaped itself from a declining industrial town to a city of tourism and services. Physical redevelopment, however, did not reverse the loss of steady, good-paying jobs, nor did it provide a secure economic future for the city.

By 1980, as a consequence of years of white out-migration and African American in-migration, the city had an African American majority. In 1987, the city's politics reflected this changed demography. Kurt Schmoke became the first African American elected to the office of mayor, and African Americans held half of the seats on the city council.

Baltimore's most recent economic development strategy, unveiled by the city's leading business association, the GBC (Greater Baltimore Committee), in 1991, focuses on the "life sciences." Advocates maintain that, although high technology drives the life sciences, an economy based in the life sciences will provide jobs that draw on a broad range of interests and skills. According to a GBC (1992) report, "A Marine Biology Center requires everything from marine biologists, to maintenance engineers, to accountants, to community education specialists, to specially-trained plumbers" (p. 5).

The "life sciences" focus springs from the region's early efforts in medicine and biological research. Johns Hopkins University, Baltimore's largest private employer, is a leader in life sciences research. The area's two University of Maryland campuses also have growing medical and science research facilities. The GBC reports that 16 of the region's top 50 employers are in health-related fields. Civic leaders have identified education and training as the "most important ingredient to achieving the life sciences vision" (p. 7). Indeed, the special skill requirements of a life science economy heighten the need for an expanded human investment policy.

In an effort to ensure that Baltimore's African American community "fully buys into, actively participates in and aggressively takes ownership" of the life sciences vision (GBC, 1992, p. 7), business and civic leaders have committed to working for increased funding for the public schools in Baltimore to bring them up to the level of the rest of the region. Business leaders have also encouraged new approaches to education, endorsing such efforts as school-based management.

Along these lines, business leaders are exploring the creation of new institutional mechanisms to carry out their commitment to ensure meaningful participation from the entire community. The GBC is working with individual African American churches and the American Association for the Advancement of Science to further the Black Church Project in Baltimore. This project promotes science training in church-based education programs. It focuses on developing enrichment programs for hands-on science, mathematics, and computer activities; trains church- and community-based organizations to start or expand existing science programs; and helps churches form networks to encourage local support for science-based education and programs. Several African American churches in southeast Baltimore are already piloting the Black Church Project. The GBC has formed a committee of prominent African American clergy, business executives, and government officials to study the possibility of institutionalizing the church project and to explore other ways of getting African Americans involved in the life sciences (Keller, 1992). In addition, the Community College of Baltimore, with support from the city's Office of Employment Development, has established a Life Sciences Center aimed at preparing city residents for a number of technical jobs.

Despite extensive redevelopment in the Inner Harbor and the identification of new economic goals, industrial decline has been very hard on the city. A shrunken tax base and population loss, especially a diminished middle class, left the city in a fiscal crunch even before the recent recession added its impact. The mayor's office has sought to control overall expenditures by downsizing city government. In 1988, the city implemented a general hiring freeze, exempting only teaching positions, critical public safety jobs, and judicial positions. A few months later, over the protests of neighborhood leaders, the city closed four fire units. Downsizing has continued, and the city has carried out significant reductions in its workforce. From 1989 to 1991, the city abolished a total of 1,442 positions (Baltimore Department of Finance, 1991, pp. 15-16; Fletcher, 1992).

Along with downsizing, Mayor Schmoke has embraced privatization as both a cost-control measure and a means of counteracting entrenched bureaucracy. A first step was a city contract with a private firm to take over the management of the Baltimore arena. The city took another dramatic step in 1992 when a private company began running nine public schools. This move came after considerable disappointment over the pace of education reform. With strong backing from Mayor Schmoke and elements of the city's civic leadership, Education Alternatives Inc. (EAI), a Minneapolis-based firm, received a five-year contract to operate the nine schools. EAI uses a curriculum and approach called Tasseract. Its model calls for small student/teacher ratios, the use of computers and other technology, and heavy parental involvement.

Although the school system retains authority over assignment of all professional staff, such as teachers, EAI can recommend assignments and transfers and set final staffing levels. EAI also can hire student teachers, interns, and other employees. Seen by some as a major setback for the teachers' union, the arrangement has received criticism on two counts, both of which involve cutting costs: First, EAI has replaced experienced teacher's aides with lower-wage interns; second, EAI mainstreams too high a proportion of special education students.

The dilemma that the mayor's office faces is that, with a boost from collective bargaining by municipal unions, city jobs are relatively well paying and drain the hard-pressed budget. At the same time, cutting these positions decreases opportunities in an already worsening job market. Privatization may save the city money, but its direct impact includes the diminishment of employment opportunities (Fletcher, 1992).

In summary, some elements of Baltimore's history are encouraging; others are discouraging. As the city continues its transition into a postindustrial future, human investment has gained top billing on the local agenda. BUILD, a community-based organization with a strong foundation in African American churches, played an active role in shifting city priorities and helped awaken business to the need to contribute to a human investment approach (Orr, 1992). The mayor and city government have accorded a high priority to education and job training. In the private sector, biracial cooperation is evident in programs such as the Black Church Project. The city's successes in redevelopment demonstrate that public-private collaboration is possible. Yet a complete picture of Baltimore shows that the city faces enormous problems and does so with very limited resources. Measures such as privatization are driven as much by a desperate need to cut costs as by the desire to improve service. Indeed, some critics fear that privatization will lower the quality of city services and reduce wage levels in the process.

Economic transformation, especially when complicated by racial concentration and a history of discrimination, is a painful and socially disruptive process. A service economy requires limited economic concentration in the center and therefore facilitates outward movement. Mobility allows what Robert Reich (1991) has called the "secession of the successful."

City Context

SOCIOECONOMIC FEATURES

Population

Table 2.1 shows the changes in Baltimore's population in the postwar decades. In 1950, Baltimore's population peaked at 950,000. Since 1950, however, the city's population has declined during each census count, down to 736,000 in 1990. This table also shows the changing racial composition of the city. As Baltimore lost population, the surrounding counties gained residents. In 1950, suburban Baltimore had 455,000 residents. In 1990, it was home to 1.6 million people, 86% of whom were white.

The data indicate that much of Baltimore's population loss can be explained by the exodus of whites from the city. The proportion of the Baltimore population that is African American rose from 24% in 1950 to 59% by 1990, whereas the white population declined from 76% to 39%. Over the 40-year span, African Americans increased in absolute numbers as well, and the white population decreased to less than half its 1950 size.

Employment and the Local Economy

Accompanying these demographic changes have been shifts in the local economy. Table 2.2 shows employment in four areas from 1950 to

TABLE 2.1 Population of Baltimore, 1950-1990 (in thousands)

Year	Total	% Change	White	% Total	Black	% Total
1950	950		724	76	225	24
1960	939	−1	611	65	326	35
1970	906	−4	480	53	420	46
1980	787	−13	347	44	431	55
1990	736	−6	288	39	436	59

SOURCE: U.S. Bureau of the Census (1950a, 1960a, 1970a, 1980a, 1990).

TABLE 2.2 Jobs in Baltimore, by Sector, 1950-1988

Job Sector	1950	1960	1970	1980	1988
Manufacturing	119,591	112,069	101,126	69,516	44,550
Wholesale and retail trade	95,981	93,059	104,415	92,882	93,905
Finance, insurance, real estate	22,286	27,799	33,394	37,921	44,763
Other services	30,593	33,079	48,882	95,902	138,100

SOURCE: U.S. Bureau of the Census (1950b, 1960b, 1970b, 1980b, 1988).

1988: manufacturing; wholesale and retail trade; finance, insurance, and real estate; and other services. Two trends stand out. First, the number of manufacturing jobs located in the city has declined significantly. In 1950, 119,591 manufacturing jobs were in the city; by 1988 that figure had declined to 44,550—a 63% drop. Second, the growth in Baltimore's economy has taken place primarily in the service sector. These jobs have more than doubled just since 1970.

Table 2.3 compares total employment in Baltimore with its surrounding suburbs. The number of jobs in the city stood at 325,000 in 1950, compared with 86,167 for its suburbs. Over this time span, the number of jobs in suburban Baltimore increased such that by 1988 more people worked in the suburbs than in the city. The number of jobs in the city increased, but only by 125,000 in four decades. During the same period, suburban jobs grew by more than 500,000—a sevenfold increase.

Given these economic trends, it is not surprising that the city of Baltimore has experienced more unemployment than any other jurisdiction in the region. Let us turn now to discussion of how this plays out

TABLE 2.3 Total Employment in Baltimore and Suburbs, 1950-1988

	1950	1960	1970	1980	1988
Baltimore	325,630	327,780	362,016	444,532	451,155
Suburbs	86,167	150,398	263,725	483,383	623,702

SOURCE: U.S. Bureau of the Census (1950b, 1960b, 1970b, 1980b, 1988).

in selected social problem indicators, starting with figures on city-suburb disparities in income.

Selected Social Problem Indicators

Median income in the city of Baltimore is well below the state average, and is less than half the amount of the state's wealthiest counties. Table 2.4 provides instructive figures from the 1990 census. The median family income for the city is $28,217—more than $10,000 under the state average of $39,386. Compare Baltimore County, a jurisdiction containing the city's oldest and most immediate suburban neighborhoods. Its median income, at $44,502, exceeds the state average and is 58% higher than the city's median income. Baltimore's more distant suburb, Howard County, and the District of Columbia suburb of Montgomery County provide an even sharper contrast; they have median incomes more than twice that of the city of Baltimore.

Even more dramatic are the poverty figures. More than one in six families in the city fall below the poverty level—three times the state average. The three suburban counties have only a tiny portion of the poverty found in the city. As is the case nationwide, the poverty rate among children is much greater than that among adults. Nearly one-third of Baltimore's school-age children are living below the poverty line. For the state, the proportion is about one in ten. Again, the three suburban counties fall well below the state average and even further below Baltimore. Significantly, income and the incidence of poverty are closely associated with level of education. For example, the proportion of college graduates in Howard and Montgomery Counties is three times that of the city.

Behind the figures on income and education is another story, a story of weak economic self-sufficiency. Currently, more than one in seven Baltimore residents are AFDC recipients. In 1992, the city reported 39,737 cases covering a total population of more than 112,000 (76,850 children and 35,682 adults). With overall populations similar in size to

TABLE 2.4 Socioeconomic Data on Baltimore and Selected Other
Jurisdictions

	Median Family Income ($)	*% Families With Income Below Poverty Level*	*% Children 5-17 Years Living in Poverty Households*	*% Persons 25+ Years With College Degree*
Maryland state average	39,386	6.0	10.5	26.5
Baltimore	28,217	17.8	31.3	15.5
Baltimore County	44,502	3.8	6.6	25.0
Howard County	61,088	2.2	3.5	46.9
Montgomery County	61,988	2.7	4.9	49.9

SOURCE: U.S. Bureau of the Census (1990).

Baltimore, Baltimore County and Montgomery County have only a
fraction of the social welfare recipients that the city has.

Contributing to economic dependency in the city is a high level of teen
births. Table 2.5 shows that nearly 1 in 10 girls in the 15 to 17 age range
(96.8 per 1,000) gives birth. This is three times the national average
(Lewin, 1992), and five or more times greater than the rates for Baltimore County, Howard County, and Montgomery County. Despite various efforts to lower that figure, it has held constant for several years.[1]

Crime is yet another indicator of the depth of social problems faced
by Baltimore. In 1989, the city's rate of 97 serious crimes per 1,000
population was double that of the region (Baltimore Department of
Finance, 1991, p. 9). Since then, the figure has gone up even further,
leading to citizen complaints that the issue has reached crisis propor-
tions. Even in the face of the city's budget squeeze, citizens have
continued to press for increased patrols (Simon & James, 1992). Especially alarming is the extent to which Baltimore's youth are exposed to
violence. A survey of 12- to 18-year-olds published in 1992 revealed
that more than half knew a murder victim, and nearly one in four had
witnessed a murder (Marbella, 1992).

LOCAL GOVERNMENT STRUCTURE

We turn now to an assessment of Baltimore's capacity to respond to
a changing economy and the social problems that accompany economic

TABLE 2.5 Adolescent Births, 1989[a]

	Baltimore	Jurisdiction of Residence		
		Baltimore County	Howard County	Montgomery County
Births to mothers under 15 years old	119	14	3	17
Rate/1,000 females under 15	5.3	0.1	0.6	0.8
Births to mothers 15-17	1,253	219	39	182
Rate/1,000 females 15-17	96.8	19.8	11.8	14.8
Total births to mothers under 18	1,372	233	42	199

SOURCE: Governor's Council on Adolescent Pregnancy (n.d.).
a. Latest year for which data are available.

restructuring. We begin with a brief description of the city's government structure. We then examine the city's budget situation and conclude with an assessment of the city's political position within the state.

In Maryland, the basic unit of local government is the county, and the city of Baltimore is treated legally as a counterpart to the state's 23 counties. The city of Baltimore and Baltimore County are distinct geographic and legal entities; each with its own taxing power, and each responsible for public services within its boundaries. Although the state's reliance on these large units of local government makes the metropolitan region less splintered governmentally than some of its counterparts in other states, the metropolitan area is still a fragmented one. There is no overarching regional government, and each local jurisdiction is separate financially from all of the others.

The mayor, Board of Estimate, and city council share formal authority. The mayor controls all city agencies administratively and also sets the budget. Baltimore is one of only a few large U.S. cities where the education system is a part of the city administration. The mayor appoints the nine-member school board, and the school budget is part of the city budget.

The mayor's power largely rests on mayoral control of the five-member Board of Estimate. This board's power emanates from its responsibility for formulating, determining, and executing the fiscal policy of the city. Thus, in addition to setting the property tax rates and operating costs for all city departments and agencies, the board approves all contracts for capital improvements, services, equipment, and supplies for the city.

The mayor appoints two of the board's members, the city solicitor and director of public works, and the mayor also sits on the board. The city is divided into six council districts, each of which elects three councillors. The city council passes ordinances, reviews mayoral appointments, and has final authority over the city budget. The council, however, can only lower, not raise, budget items. The council is presided over by its president, who is elected by the city as a whole. The mayor's domination of citywide affairs is largely unchallenged, however, and city council members are often concerned mainly with parochial matters relating to their constituents.

City Revenues and Expenditures

Figures 2.1 and 2.2 display the operating revenues and expenditures for the city of Baltimore as of 1993. Baltimore's primary money source is the property tax, which in 1993 accounted for 25.5% of the city's revenues. State grants are the second-largest source, providing 21% of the city's operating revenue in 1993. Although local governments in Maryland are allowed to "piggyback" on the state's income tax, they can tax residents only. Thus the city cannot tax those who work in the city but live in the suburbs. In addition, because the level of the local income tax is a factor in residential location, the city is substantially limited in its ability to use the piggyback as a source of tax money. The income tax provides only 6.5% of the city's revenue.

Figure 2.2 shows how Baltimore's revenues are used, with 42% going to public education. Public safety accounts for 16%. No other category of expenditure comes close to these figures. Most social welfare expenditures are covered by the state.

Even though Baltimore's budget has been tightly constricted, the city remains financially sound. Its bond rating, however, is a modest A1/A, and it has a long-term debt per capita of $510. Thus, although the city faces no acute financial crisis, it is significantly constrained by its economic situation.

Political Position

Population decline has meant decreasing electoral influence. In 1920, half of the people of Maryland lived inside the Baltimore city limits. In 1950, when the city's population first began its decline, the figure was still at 40%. By 1990, however, Baltimore constituted only 15% of the

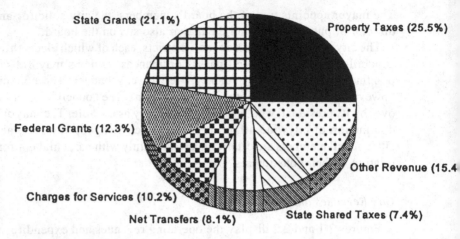

Figure 2.1. General Operating Revenues, City of Baltimore, 1993

state's population. Baltimore's voting power in statewide elections and representation in the state legislature declined accordingly. For example, in the 1948 presidential election, 43% of the total state vote was cast in the city of Baltimore. By 1988 that figure diminished to 14%, less than a third of what it had been.

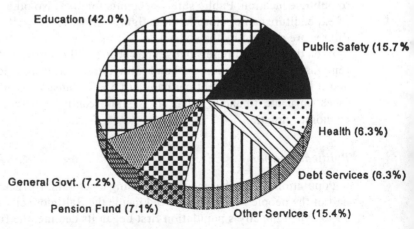

Figure 2.2. Expenditures, City of Baltimore, 1993

As the state's only large city, Baltimore lacks ready-made allies in the state legislature. Particularly frustrating has been the city's inability to achieve enough state aid to equalize school spending. Even though a suit filed in 1978 was unsuccessful, continuing legislative resistance to school finance equity has led city officials to consider a new court action. Suburban representatives have greeted the possibility of litigation with the charge that such a move would polarize the legislature and make future action even more difficult (Frece, 1992).

SUMMARY

Economic change, population loss, and decreasing electoral influence have been accompanied by worrisome trends in poverty, drug traffic and crime, teen pregnancies, and unemployment. Although Baltimore continues to be vital to the economic and cultural life of the state, the city faces intensifying social problems at the same time its resources are diminishing. Moreover, these trends have the potential to aggravate racial tensions, because the city is majorly African American and the suburbs are predominantly white. In the state and the metropolitan area, the potential for "us against them" thinking is great, even though the fundamental reality is one of interdependence.

Central-City Renewal Through
Human Investment Programs

A popular myth holds that urban human investment programs—education and job training specifically—have been amply funded and that continuing social ills are evidence that more money is not a solution to the problems that cities face. Such thinking implies that further human investment is not an effective form of spending. The reality is different. Many human investment programs do work, but they are poorly funded. Cities lack the resources to do all that could and should be done. Consequently, many needs go unmet, in some cases even when money spent in the short run would save expenses over the long run.

City schools are asked to do more with less money than their suburban counterparts. Similarly, federal officials ask job training programs to move people into economic self-sufficiency at rates that are unrealistic given the resources provided and the rules imposed. It is time to correct the misunderstanding that the federal government provides cities with generous assistance for human investment needs. In the remainder of this report, we attempt to replace the myth of federal generosity with a realistic picture by using the Baltimore case to illustrate the sparseness of federal aid in relation to city needs.

PRESCHOOL PROGRAMS IN BALTIMORE

Head Start and related early childhood programs are acknowledged successes of the Great Society (Zigler & Muenchow, 1992). Experts agree that early intervention is far more effective than later efforts at compensatory education. Yet the nation is far short of adequate funding for such programs; and cities, where need is heavily concentrated, lack the resources to do very much on their own. Consider the Baltimore situation.

The Baltimore City Child Care Resource Center estimates that in 1990, 35,511 children ages 2 to 4 lived in Baltimore. The largest percentage of this age group lived in the city's poorest areas. According to the Resource Center, in 1992 there were 139 full-day, center-based child-care programs in Baltimore with a capacity of only 6,527 children. Many of these centers are run for profit and do little to further child development.

The city of Baltimore itself currently operates four child-care centers. Mayor Schmoke, feeling the pressures of fiscal constraint, and consistent with the city's general response to its financial difficulties, in 1991 began negotiating with the City Union of Baltimore to privatize the four centers and to sever the city's financial obligation for the operation of the centers. However, issues concerning the child-care centers' employees' potential loss of health and retirement benefits have come to the fore, and the employees have made a counterproposal to bring the centers into the public school system. Moreover, privatization raises questions not only about the level of compensation that the 49 child-care center employees would receive as private sector employees, but also about the type of day care offered by private provision (e.g., developmental or mere custodial care).

In an interview, Mayor Schmoke acknowledged that fiscal considerations are an important consideration, but he also made a "reinventing government" argument—suggesting that the city's "best role is as a facilitator of child care, not as an operator." The mayor suggested that the city could sponsor child-care conferences and facilitate the creation of more child-care providers by conducting training seminars and helping potential providers to obtain funding. What this "reinvented" public role in child care does not confront is the fact that the city by itself has little leverage over an industry that is market based. In the United States, work with children in general is low paying, and child-care work is notoriously so. City wages in this industry are even lower than those in the suburbs. Thus, in using the private market to diminish its financial obligations, city hall is putting additional workers into competition with one of the city's lowest-paying industries. Market competition and low wages work against the employment of highly skilled professionals and a noncustodial approach to day care. An emphasis on early childhood development is labor-intensive and therefore not something encouraged by market-driven cost consciousness. Although the city operates only four centers, serving 334 children, three-fourths of these children are AFDC recipients and especially in need of development rather than mere custodial care. What the future holds is unclear, but a reliance on private centers does not bode well for AFDC children.

Head Start meets some of the need for developmentally oriented child care, but only a small part. The majority of Baltimore's Head Start programs are half-day care. Currently enrolling 2,560 students, Head Start operates in 29 sites around the city. Officials in Baltimore estimate that they reach only 30% of the 3- to 4-year-olds whose parents' income levels qualify them for Head Start.

Local experts indicate that not only is there a need for more full-day centers, but also that some areas need wraparound care (care offered before and after school hours) and more family child-care providers. Yet the shortage of city funds is decreasing rather than increasing the city's role, especially in providing developmentally appropriate preschool programs. Because federal and state funds do not fill the gap, reliance on the private market serves only to perpetuate a low-wage industry that is insufficiently concerned with developmental aims.

CITY/SUBURB DISPARITIES IN EDUCATION

Education is central to any effort to improve the level of economic self-sufficiency among the city's population. Yet education is an arena in which disparities between the city and the suburbs are especially high. To a greater extent than in any other advanced industrial nation, the United States relies on local financing for education, that is, for elementary and secondary education. And it is precisely at these levels (not in higher education) that the United States falls behind many other advanced nations.

The state of Maryland provides more than half of the funding for the city's school system. The federal government contributes only 12%, and the local share is 33%. This approximate breakdown of local, state, and federal funding for Baltimore city schools has held for several years, despite the city's plea for greater assistance from the state.

There is some equalization money, and Baltimore's schools do receive more state and federal money than do the suburban jurisdictions. From that pattern one might suppose that education spending has been equalized, but, as Table 2.6 shows, disparities continue—even though Baltimore has acute human investment needs. Thus, despite some state and federal assistance, Baltimore spends considerably less per pupil than the state average, whereas the three suburban jurisdictions spend more.

The disparities are not based on Baltimore's unwillingness to tax. The *Preliminary Strategic Financial Plan for the City of Baltimore* reports that the property tax rate in the city is *double* that of most jurisdictions in the state (Baltimore Department of Finance, 1991). As Table 2.6 makes clear, on a per pupil basis, Baltimore has a smaller tax base than the state average and even smaller in relation to the suburban counties. Per pupil wealth in the city of Baltimore is roughly half that of Baltimore and Howard Counties and one-third that of Montgomery County. Once the extreme differences in per pupil wealth are considered, it is not surprising that differences in per pupil expenditure persist in the face of present levels of state and federal assistance and in the face of the city's high property tax rate.

Disparities in per pupil expenditure assume even greater significance when student needs are taken into account. Because family background confers significant educational advantages on some students and disadvantages on others, the concentration of poverty in the city poses a greater challenge to the Baltimore school system than to its suburban counterparts. The specific figures are telling. The dropout rate in the city is more than three times the state average.

If the estimated 5,000 students who have dropped out of Baltimore's schools at the rate beyond the level for the rest of the state were included in the per pupil calculations on spending and in the wealth per pupil calculations, then Baltimore would fare even worse in comparison with

TABLE 2.6 Education Data for Baltimore and Selected Other Jurisdictions

	Maryland State Average	Baltimore	Baltimore County	Howard County	Montgomery County
Per pupil expenditure ($)	5,815	4,947	6,220	6,695	7,591
Per pupil wealth ($)	210,777	121,950	260,563	259,359	360,629
Instructional staff per 1,000 students	60.7	57.4	64.5	64.0	62.8
Professional support per 1,000 students	9.1	7.2	10.7	10.9	10.4
Instructional assistants per 1,000 students	9.4	9.2	5.0	13.3	13.6
Average teacher salary ($)	38,685	37,039	39,378	39,364	45,393
Special education students (%)	11.4	14.5	11.8	10.7	9.6
Chapter 1 students (%)	9.0	24.1	8.9	2.8	5.3
Students assisted with school lunch (%)	26.1	67.0	14.7	5.9	17.0
Limited English proficient students (%)	1.6	0.3	1.1	1.4	5.6
Dropout rate (%)	5.2	16.4	3.0	2.0	2.3
Students attending college (%)	58.9	50.6	58.5	77.5	75.5

SOURCE: Maryland Department of Education (1992).

other jurisdictions in the state on both of these measures. In other words, education expenditures per school-age child and wealth per school-age child are overstated by figures based on school enrollment.

Dropout figures parallel indicators of student background. Two-thirds of the students in Baltimore's public schools are eligible for federally assisted school lunches. The suburban figures are only a small portion of that percentage. Similarly, a quarter of the city's pupils qualify for Chapter 1 assistance from the federal government, whereas the proportion in the suburbs is far smaller, falling to less than 3% in Howard County. Even the special education responsibilities of the city are greater than those of suburban jurisdictions. Only on limited English proficiency does the city face little challenge.

Having less money to spend on education has concrete consequences. Despite its substantially greater needs, the city of Baltimore has fewer school staff in relation to number of students than the state average, and even fewer in relation to the suburban average. City teachers are also less experienced on the average, and less well paid.

The cumulative enormity of the disparity between city and suburbs in education spending can be lost among its many particulars. But it can perhaps be conveyed in the following exercise: For a classroom of 30 students, the discrepancy in per pupil expenditure between the city of Baltimore and Montgomery County amounts to nearly $80,000 a year. Thus a classroom with a large number of needy students is substantially outspent year by year by a classroom with far fewer needy students. Over the 13 years represented by kindergarten through high school, this amounts to more than a million dollars.

THE TRANSITION FROM HIGH SCHOOL

Originating with pressure from BUILD, a community-based organization instrumental in shifting city priorities to human investment issues, the Baltimore Commonwealth is a compact agreement with the Greater Baltimore Committee (representing the business community) to provide assistance to high school students as they enter the job market or pursue opportunities in higher education (Orr, 1992). Under Kurt Schmoke, the Commonwealth arrangement has been incorporated into the city government as part of the mayor's effort to make the most of the partnership with the GBC.

The Baltimore Commonwealth offers a "guarantee of opportunity" for all students in the Baltimore public schools who maintain 95% attendance records during their junior and senior years. Each high school graduate with a good attendance record is guaranteed three job interviews at GBC member companies, and, if the interviews themselves do not produce a job, the city Office of Employment Development evaluates the young person and provides additional training.

Another important feature of the Commonwealth is the College-Bound Foundation. Formed in the spring of 1988, it is an endowment collected by the Greater Baltimore Committee to provide students with assistance to attend college. Baltimore's business community pledged to raise $25 million for the foundation. As of this writing, some $13 million has been collected.

The CollegeBound Foundation (1992) says that it "works with students to encourage them to take the SATs, assist in college selection, fill out complicated financial aid forms and award deserving students last-dollar financing" (p. 1). The program attempts to increase the awareness of students and parents of the availability of financial support to attend college. Through CollegeBound, paid advisers go into the high schools to help students learn of scholarship opportunities and to assist them with the often overwhelming application process. During the 1990-1991 school year, CollegeBound advisers made presentations to nearly 9,000 high school students through group meetings. More than 2,500 one-on-one sessions were conducted by advisers with students to discuss college options, courses needed, SAT registration, and financial aid applications.

WELFARE-TO-WORK TRAINING EFFORTS

Like most urban areas in the United States, Baltimore has a large number of adults without work and without the skills to secure employment. Adults receiving public assistance are in particular need of help, not only because they are a drain on precious tax dollars, but because they are usually parents. Given the sheer magnitude of families receiving Aid to Families with Dependent Children (AFDC) who reside in central cities, any effort to achieve central-city renewal via a human investment strategy is likely to flounder without significant progress on this front. Upgrading the education levels and job skills of persons receiving public assistance is clearly an important element of the human capital approach to urban revitalization, and the federally mandated Job Opportunities and Basic Skills (JOBS) training program is intended to address this challenge.

With the passage of a major piece of welfare reform legislation in 1988—the Family Support Act (FSA)—the federal government embarked on a nationwide attempt to redress the chronic human capital deficiencies among welfare recipients. At the core of the FSA legislation is the JOBS program, a new employment training program providing matching federal funds for states to create, within the bounds of federal specifications and regulations, initiatives designed to prepare these persons for the labor market and economic self-sufficiency. Unlike previous federal employment programs for AFDC recipients, the JOBS legislation emphasizes job readiness strategies involving intensive, long-term human investment interventions for the most disadvantaged

AFDC clients. However, the experience in Baltimore shows that local officials have been forced to implement the program in ways that undercut legislative intent.

Prior to the beginning of the JOBS program, Baltimore amassed a long and distinguished record of effectively providing job training assistance for the disadvantaged. Its Office of Manpower Resources (now the Office of Employment Development) was "among the most experienced and highly regarded CETA/JTPA operators in the country" (Gueron & Pauly, 1991, p. 176). Baltimore quickly responded when the federal government first expanded the authority of states to develop improved employment programs for AFDC registrants in the early 1980s. Whereas most other pre-JOBS welfare-to-work initiatives offered less intensive services designed only to move people quickly off the welfare rolls, the Baltimore pilot program (called Options) put the "emphasis on enhancing participants' long-term employability through involvement in education and training. . . . [Helping] individuals . . . find jobs that will confer at least a modicum of economic security" (Friedlander, Hoerz, Long, & Quint, 1985, p. 6). The results of a comprehensive evaluation showed that this approach produced positive results, especially over the longer term: By the second and third year after enrollment, program participants earned 17% more than similar AFDC recipients who did not participate (Friedlander, 1987; Gueron & Pauly, 1991, p. 17).

In light of this history, and given the newly elected mayor's top priority of increasing human investment efforts in the city, Baltimore's job training and social service officials "hailed the change in federal philosophy that was articulated in the [FSA] legislation, believing, in fact, that this was the opportunity to implement a local strategy that fostered and supported the movement of welfare recipients into the labor market mainstream" (Harris, 1991, p. 58). The city, in conjunction with the state of Maryland, engaged in a highly proactive and inclusive planning process to prepare for the implementation of the new law. These measures, along with the presence of a delivery system staff that was "highly energized and excited" about the new program, allowed the city to develop a comprehensive and innovative JOBS program (Harris, 1991, p. 58). This program was targeted toward the hardest-to-serve cases and emphasized intensive human investment strategies—hence faithfully fulfilling the federal legislative objectives.

But when the Department of Health and Human Services (HHS) issued regulations for the program, Baltimore officials realized their intensive, targeted program could not be sustained given the very high number of recipients that federal authorities expected local agencies to assist. Although the legislation *did* call for a high-volume program (by requiring states to meet a monthly participation rate that will increase to 20% of the nonexempt caseload by 1995), the regulations made this aspect of the law especially troublesome in two ways. First, regulations required that a group of clients must spend an average of 20 hours a week in training (with a 75% rate of attendance) to be counted as participating in the program (the so-called 20-hour rule). Second, regulations required local programs to use a complicated methodology for calculating participation rates that is "fundamentally irrational" and in practice works to undercount severely the actual level of client participation in the program.[2] As a consequence of these regulations, Baltimore officials found that, in order to achieve the required number of "countable" participants, they must in actuality serve considerably more clients than the number mandated by the law's participation rate standards.[3]

This situation has been further exacerbated by the inadequate level of funding provided for the JOBS program. From 1991 to 1992, for example, statutory participation rates increased by more than 50%, whereas the level of federal expenditures for the program remained constant. In addition, the recent recession has taken its toll. Not only has it caused a dramatic rise in the size of the AFDC caseload—and thus a corresponding increase in the number of clients who must be served by JOBS, it has prevented cash-starved states (such as Maryland) from allocating enough of their own funds to the program to receive their entire share of federal matching dollars.[4]

Overall, as a result of these fiscal constraints and burdensome regulations, to serve the required number of clients Baltimore officials were forced to dismantle much of their original JOBS program. Intensive human investment services were replaced with short-term, low-cost activities that serve substantially more participants. Local implementers also were compelled to recruit a pool of clients who are less in need of substantial remediation to become job ready, and thus less expensive to serve. Thus, compared with the initial Baltimore version of JOBS (which was judged to be very successful by an evaluation study), the current, revamped version is driven more by a concern for putting enough bodies through the system to meet federal participation rates than by the need to provide the kinds of in-depth education and training that would propel AFDC recipients toward economic self-sufficiency.

The other major problem local implementers in Baltimore have faced is inadequacy of resources. Much of the difficulty here is linked to the structure of the funding mechanism for the JOBS program. In order to receive federal matching dollars, states must devote substantial amounts of their own funds to the program. Like many other states, the budgetary crisis caused by the recent economic slowdown prevented Maryland from appropriating the necessary funding to draw down its entire federal match. This tended to reinforce the effects of the recession, as states were denied resources to operate their JOBS programs just when they needed those resources most.

Finally, a general work incentive problem continues to hamper the city's efforts to move clients from training programs to actual employment. Clients making this move normally lose their AFDC eligibility, but to ease the impact of this transition the FSA legislation allows them to continue to receive Medicaid and child-care assistance for up to a year after entering the workforce. However, former welfare recipients taking jobs still see their food stamp benefits decrease and, if they are residents of public housing, their rents rise. And, given the city's depressed urban labor market, employed former clients are likely to earn low wages, especially if they have not received the kind of intensive employment training services the Baltimore JOBS program is increasingly unable to provide. Thus entrance into the workforce for many clients may entail a substantial sacrifice in real income. As a result, Baltimore officials report that it is more difficult to move welfare recipients off the AFDC rolls and into jobs.

Effects of Programs

Government-initiated or -sponsored human capital investment programs in Baltimore provide no miraculous cures for central-city decline. The main story to be told, however, is not one of the futility of city efforts. Quite the contrary, pilot projects such as Baltimore's Options program demonstrate an ability to make a difference. That small successes have not been converted into big ones is largely a matter of inadequate funding and cumbersome federal regulations. Although our coverage of human investment programs is not exhaustive, it does provide a picture of the range of Baltimore's efforts.

BALTIMORE COMMONWEALTH AND COLLEGEBOUND

CollegeBound officials point to the increase in the percentage of Baltimore public high school students taking the SATs as an example of the program's success. During the 1990-1991 school year, the percentage of students taking the examinations went up 24% over the previous academic year. This increase is especially impressive because there was a slight decrease statewide. During the same period, the proportion of seniors completing financial aid forms for assistance with college expenses rose 46%. And there was a 77% increase in the number of students who completed college applications, growing from 26% of seniors in 1990 to 46% in 1991 (CollegeBound Foundation, 1992).

The foundation also provided students with grants to help bridge the gap between conventional financial aid sources, such as scholarships and school loans, and the ability of students to meet all of their needs. At this writing, there are 167 college students receiving CollegeBound "last-dollar grants." All grant recipients are 1989, 1990, and 1991 graduates of the city's high schools. The average CollegeBound grant was $1,137, with the total amount of money awarded during the three-year period reaching almost $190,000.

The Commonwealth also has summer youth employment programs whereby Commonwealth students work in city agencies, community-based organizations, and private businesses. In 1990 (the most recent year for which figures are available), the Commonwealth helped more than 3,600 teens to secure summer employment in 460 community-based organizations and city agencies as well as more than 100 private businesses.

Although the Commonwealth and CollegeBound programs are laudable local initiatives, they are not enough. The national recession caused many locally based companies to retreat inward, and to pay less attention to community problems. The admittedly slow pace of the business community to raise the promised $25 million for the CollegeBound Foundation is attributed to the recession and changes in corporate management personnel.

In 1992, the demand for CollegeBound Foundation last-dollar grants increased so that for the first time in its four-year history demand outstripped supply. An unexpectedly high number of additional students, about 24, qualified for last-dollar grants for 1992. According to CollegeBound's executive director, these students faced the possibility of not having funding to go to college. A Ford Foundation grant funneled through the Baltimore Community Foundation provided a little breathing room. Nevertheless, the Baltimore experience shows that private money is insufficient to do more than provide supplementary assistance, and even that is easily exhausted by increases in the number of students seeking to enter college.

CHILD CARE

Privatization initiatives in the area of child care could result in a decrease in the salaries of a small number of child-care workers and a diminishing of the developmental aspects of these programs. In the short run, privatization saves the city a small amount of money. In the long run, however, cutting preschool programs creates potential liabilities for the future. The city can promote discussion of child-care goals and methods, but by itself the city has little leverage over the performance of the industry—whether it maintains its four centers or privatizes them. An appropriate child development policy is a larger task than can be handled at the local level.

JOB TRAINING

The effects of the changes in the JOBS program also reveal the limitations under which local governments operate. Restrictive federal regulations have resulted in short, ineffective training that does little or nothing to improve the skills of trainees. The ability of Baltimore to target those most in need of assistance has been severely impaired, and the morale of city workers has been damaged. In essence, federal regulations forced Baltimore to run an employment program for job trainers, not trainees. The effects of the rules regarding the transition from public assistance to self-sufficiency are particularly ironic, as they have provided a disincentive to those most willing and able to work. Without a change in federal regulations, the JOBS program will expend tax dollars with too little return on the investment. As with the child-care system, the losses are primarily in opportunity costs. If participants in the JOBS program had received the intensive training planned and demonstrated by the city instead of the short-term training mandated by HHS regulations, Baltimore would have fewer people on the welfare rolls and more paying taxes.

Policy Implications

RECOMMENDATIONS:
HEAD START TO HIGH SCHOOL

Although money is not the total solution to the problems of urban education, it is a necessary element in any effective response. Cities are asked to do more than their suburban counterparts and to do so with fewer resources. Consequently, there are significant gaps in the urban human investment effort.

An examination of the Baltimore situation reveals that Head Start and appropriate forms of child care fall far short of full coverage. Insufficient early childhood programs are followed by inadequate spending in the Baltimore public schools. The transition from high school to college or work is now aided by private assistance through the Baltimore Commonwealth, including the CollegeBound program. The success of the Commonwealth in stimulating college applications, however, exposes a weakness of the program. As interest in college attendance grows, the inadequacy of private funds becomes evident. More money is needed, then, at all levels—preschool, K-12, and postsecondary. Because Baltimore already taxes at a high level, the solution is not local revenue. The city needs more help than it is now getting from the state and federal governments. Our recommendations are as follows:

1. Federal funding should be sufficient for all eligible children to enroll in Head Start.

2. Because half-day Head Start is inadequate to meet the needs of many poor families, preschool subsidies should enable Baltimore and other cities to provide flexible forms of day care that "wrap around" Head Start and other early childhood programs and accommodate both the schedules of working parents and the development needs of children.

3. The Commission on Chapter 1, composed of school officials, child advocates, and business executives, has recommended concentrating federal money for education where there is more poverty (Jordan, 1992).[5] We endorse this proposal to increase federal education assistance to cities. It is likely to take both greater state and greater federal aid to move city schools closer to the level of spending that suburban systems enjoy.

4. Baltimore demonstrates that private resources can be used to stimulate greater interest by inner-city students in college attendance. Advising makes a difference, as does covering application and SAT costs for needy

students. Private money, however, is insufficient to fill the scholarship gap for the poorest students. As the nation considers changes in the college loan program and the possibility of loan "forgiveness" through a national service obligation, it is important that adequate assistance for the neediest students not be overlooked.

RECOMMENDATIONS: JOB TRAINING

The Baltimore implementation experience with welfare-to-work training programs shows that, if policymakers are serious about building a genuine human investment apparatus for welfare recipients, it will be necessary to do at least the following:

1. Revise the calculation methodology so that it accurately represents the actual level of client participation in the program. In addition, assign special weight to the participation of those in harder-to-serve groups and reward serving these clients, who need intensive, expensive services to become job ready.

2. Address the remaining work incentive problems in the income maintenance system. The transition to economic self-sufficiency can be hampered if employment results in a sudden loss of benefits. Under current policy, AFDC clients finding jobs are eligible for transition assistance in health and in child care, but not in public housing and nutrition assistance. Extending transition benefits to these program areas would make it easier for AFDC recipients to move into the world of work.[6]

Conclusion

As a case study in central-city revitalization, Baltimore provides a number of lessons and insights. First of all, Baltimore is a clear example of the shift from an industrial to a service economy. Revitalization efforts must confront that fundamental fact. Second, adjustment to basic economic change cannot be accomplished effectively simply by rearranging land use and putting up new buildings. A human investment strategy is a necessary element in any transition to a rejuvenated economic foundation. Baltimore, for instance, has achieved a highly celebrated redevelopment of its Inner Harbor, yet the city continues to struggle with major social problems, most of which are rooted in the incomplete incorporation of the city's population into the mainstream economy.

Now that the national discourse on economic policy is greatly concerned with human investment issues, it is important that the special circumstances of older industrial cities such as Baltimore be examined carefully. From that examination, we can draw a third lesson—namely, that the human investment challenge faced by such cities is extraordinary. Urban needs are increasing at the same time city resources are diminishing.

Some might argue that the central city should simply be abandoned, that it should give way to new "edge cities" (Garreau, 1991). Such a view, however, is callous, inequitable, and shortsighted. It is callous because the well-being of many people is tied up in the central city. It is inequitable because government action and inaction helped stimulate selective outward movement from the central city, thus contributing to the problems that cities face. Finally, abandoning the central city would be shortsighted because it remains the location of many jobs and considerable sunk investment. On all three grounds, federal and state governments would be well advised to promote human investment in such a way as to lessen the disparities that separate urban from suburban citizens.

If the cities are to become targets of a human investment strategy, then several misunderstandings need to be cleared up. The main one is that social programs are adequately funded but don't work. The opposite is the case. For example, programs such as Head Start perform but are *greatly* underfunded. Further, despite talk about equal opportunity, the local financing of education creates huge inequities. The disparities between city and suburban expenditures on education are enormous. Without substantial federal and state expenditures, inequities will continue. To carry the argument a step further, there is no rationale for disparities in education spending. The motivation of the poor to achieve educationally and move ahead economically is abundantly clear when favorable circumstances prevail, as under Baltimore's Commonwealth program. Human investment receives ample lip service in national legislation, as in the Family Support Act, but the actuality is quite different. Whether intended or not, federal regulations have worked against a human investment approach to welfare clients.

Movement of welfare recipients from economic dependency to self-sufficiency has to be planned and executed carefully. As the nation's industrial economy has declined, steady work that pays a decent wage is not as readily available as it once was. During the industrial era, factory work required few entry-level skills. Then, the business cycle was the major threat to economic self-sufficiency, and unemployment benefits and pension rights were effective counters to economic insecurity. In the present postindustrial era, characterized as it is by the growth of contract work and other forms of temporary and part-time employment, the nature of the job market itself is undergoing change. Much more is involved than the ups and downs of the business cycle.

A national policy of human investment needs to recognize the extraordinary problems cities face. Such a policy should also take into account the changing nature of the job market as the economy moves from an industrial to a postindustrial base. In recent years, Baltimore's government, business, and community leaders have acted on several fronts to cope with economic change. If they are to be successful in their efforts, they need a responsive federal government. This means more resources directed into cities and more appropriate federal regulations. Human investment is not a cost-free exercise. However, a failure to invest money now may be even more costly in the long run.

Notes

1. In response to the persistence of the problem, Baltimore public health officials have discussed offering Norplant (the surgically implanted contraceptive) in city schools (Lewin, 1992).

2. Baltimore officials discovered two flaws in this methodology: (a) Training programs must start during the first week of a month and end in the last week of a month in order to count the actual hours of client participation, and (b) if clients are placed in a training activity during the same month that they get their original employability assessment, the participation count is lowered (L. Harris, director, Office of Employment Development, interview, December 4, 1992).

3. To illustrate: In federal fiscal year 1992, the required monthly participation in Baltimore was 3,585 "countable" participants. To reach this figure, the Office of Employment Development had to serve 4,781 clients each month. Figures thereafter were projected to continue upward (Office of Employment Development, internal document).

4. In Baltimore, from 1989 to 1992 the number of AFDC recipients grew by almost 11,000, a 10.6% increase. Maryland received only 82% of these monies in 1991 (Harris, 1991, p. 59). All states claimed only about $600 million of the $1 billion available in that year (Eckholm, 1992, p. 18).

5. The commission also recommended more flexibility in Chapter 1 (Commission on Chapter 1, 1992; see also Rotberg & Harvey, 1993; U.S. Department of Education, 1993).

6. Current discussions of welfare reform include proposals to limit the time recipients are eligible to receive assistance, thereby emphasizing the goal of moving recipients into employment. If such measures are to be something other than punitive, the issues of supplemental benefits and adequate education and training need to be dealt with.

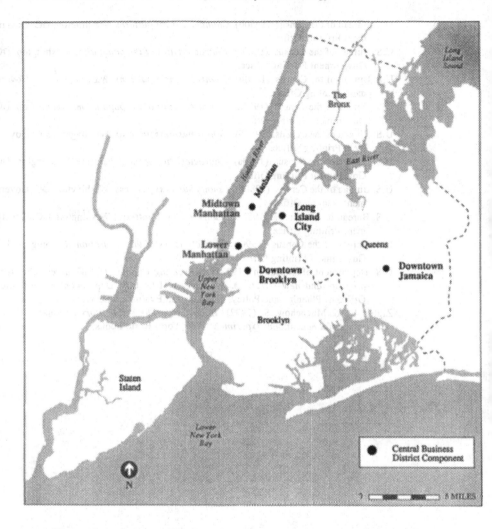

Figure 3.1. Central Business District Components, New York City
SOURCE: New York City Department of City Planning.

ment as much as civic improvement: the "search for a dignified and suitable civic center" for Brooklyn.

By the late 1960s, the continuation of racial transition, white flight, and property disinvestment trends begun in the 1940s gave rise to a new

U.S. Bureau of the Census. (1950b). *County business patterns.* Washington, DC: Government Printing Office.

U.S. Bureau of the Census. (1960a). *Characteristics of the population.* Washington, DC: Government Printing Office.

U.S. Bureau of the Census. (1960b). *County business patterns.* Washington, DC: Government Printing Office.

U.S. Bureau of the Census. (1970a). *Characteristics of the population.* Washington, DC: Government Printing Office.

U.S. Bureau of the Census. (1970b). *County business patterns.* Washington, DC: Government Printing Office.

U.S. Bureau of the Census. (1980a). *Characteristics of the population.* Washington, DC: Government Printing Office.

U.S. Bureau of the Census. (1980b). *County business patterns.* Washington, DC: Government Printing Office.

U.S. Bureau of the Census. (1988). *County business patterns.* Washington, DC: Government Printing Office.

U.S. Bureau of the Census. (1990). *Characteristics of the population.* Washington, DC: Government Printing Office.

U.S. Department of Education. (1993). *Reinventing Chapter 1: The current Chapter 1 program and new directives.* Washington, DC: U.S. Department of Education, Office of Planning and Policy, Planning and Evaluation Service.

Zigler, E., & Muenchow, S. (1992). *Head Start: The inside story of America's most successful educational experiment.* New York: Basic Books.

3

New York City's Outer Borough Development Strategy

Case Studies in Urban Revitalization

EDWARD T. ROGOWSKY
RONALD BERKMAN
with
ELIZABETH STROM
ANTHONY J. MANISCALCO

The pursuit of growth through economic development has become the key strategy in efforts to revitalize older American cities. All levels of government have come to play roles in the complex intergovernmental process of designing, funding, and implementing economic development programs. In New York City, federal funds—especially Urban Development Action Grants (UDAGs) and Community Development Block Grants (CDBGs)—have been essential components of local development policies and programs.

In this chapter we present case studies of urban revitalization in selected areas of New York City outside of Manhattan. The areas chosen for study in Brooklyn and Queens, usually obscured by the attention given to Manhattan, have been the focus of concerted efforts to promote the revitalization of commercial hubs and surrounding residential neighborhoods. In these case studies, we describe and analyze the goals, designs, implementation, and impacts of federal funding within the broader context of New York City's economic development policies and programs. The projects studied had multiple and related goals, including job creation and retention, commercial development and support for retailing activity, and the physical upgrade of deteriorated neighborhoods. The efforts described in these case studies constitute the core of the City's "outer borough development strategy."

As in other municipalities, the growth of a city's central business district (CBD)—in New York City, the area from 60th Street south to the Battery, on Manhattan Island—has historically been the key to the development and economic well-being of the entire city. "The enormous concentration of activity—40% of all employment in the City—in this relatively small geographic area of New York is the defining characteristic of the City's economy" (New York City Planning Commission, 1993, p. 28). The efforts studied were initiated by a decision to extend the borders of the traditional central business district beyond Manhattan, based on an underlying governmental assumption that for the City's economy to expand over time, new commercial centers must eventually be created (New York City Planning Commission, 1993, p. 33).

Among other themes, our work explores a dominant theory in the academic literature on urban economic development: that market forces and economic elites exert dominant influence on local policies. In studies of Atlanta and Dallas, Stone (1989) and Elkin (1987), for example, have concluded that private sector elites are far more important than public actors in shaping local economic development policy and planning. The case studies presented in this chapter provide an opportunity to assess the validity of this axiom in a more complex intergovernmental setting.

City Context

OVERVIEW

To many observers, New York City *is* the island of Manhattan, where many of the nation's most important institutions of finance and culture are situated. In fact, most New Yorkers live in one of the other four counties or boroughs—Brooklyn, the Bronx, Queens, and Staten Island—that constitute today's city. Of the five boroughs, only the Bronx is actually on the mainland. The other four boroughs are on a series of islands connected by bridges and tunnels. This geography makes infrastructure and transportation issues particularly salient. With its population of 7.3 million, New York City lies at the center of a large metropolitan area that includes suburban counties in New York State, New Jersey, and Connecticut. The population of the Greater New York Metropolitan Area approaches 20 million ("U.S. Says," 1991).

GOVERNANCE AND POLITICS

The mayor and the city council are the dominant political forces in New York City government. The city comptroller and public advocate (formerly city council president) are the two other officials elected citywide. Each borough elects a borough president. In addition, state and city commissioners and local community boards (filled by unsalaried appointees) play roles in the policy process. The mandates of city officials as well as their actual power to affect decision making and to mobilize resources have changed with successive charter reforms.

One of the most significant changes in the political landscape occurred recently when New York City was ordered to change its basic structure of government. Prior to 1990, a body called the Board of Estimate, consisting of the mayor, city comptroller, city council president, and the five borough presidents, shared power with the city council on the budget and exercised final authority on land use and contracting matters. This was an avenue through which borough presidents could exert significant influence over city policy making and protect and enhance the fortunes of their boroughs. This was the form of government in place during the period chronicled in the following case studies. However, following a U.S. Supreme Court ruling that the equal representation of the unequally populated boroughs violated the nation's one-person, one-vote principle, the Board of Estimate was abolished and a newly enlarged district-based, 51-member city council—always the city's legislative arm—took on many of the board's responsibilities in land use, planning, and other significant matters of urban revitalization.

The Democratic party has long dominated local politics. Rudolph Guiliani, who defeated David Dinkins in November 1993, is only the third

Republican mayor since World War I—each elected with considerable help from defecting Democrats. The current city council has just seven Republican members. Conflicts have thus been most frequently played out in intraparty disputes, usually in primary elections. Local general elections, with rare exception, are not the flash point of local politics. Conflicts within the Democratic party have tended to be either along racial/ethnic lines or between the "regulars," who adhere to the machine-style politics of their Tammany Hall forebears, and various kinds of "reformers," who have tended to be more liberal and ideological in their policy preferences. The persistence of machine politics in New York City, in spite of the demise of an organized political machine, means that patronage and exchange of favors as well as periodic corruption scandals are part of the local political milieu. These are elements of the culture within which development policy is forged.

Race and ethnicity represent a second, often more significant, political cleavage. Ethnic divisions in local politics are hardly new. Tammany Hall leaders took great pains to present a "balanced ticket," slating Irish, Jewish, and Italian candidates to appeal to the largest segments of the electorate. Ethnic and racial politics have come to the fore again recently as the city has experienced a decade of massive immigration (for an in-depth analysis of demographic change in New York City, see Mollenkopf, 1993). It is estimated that one-fourth of all immigrants to the United States enter through New York City. The percentage of current New Yorkers who were born abroad stands at approximately 30%, approaching the number of foreign-born found in the population of 1900, the peak year of the largest previous wave of immigration. Whereas most of the city's out-migrants have been white, those who have migrated to and remain in the city are peoples of color. As in cities such as Atlanta, Chicago, and Detroit, whites are now a minority in New York City. Unlike those other cities, however, in New York no one racial or ethnic group can claim majority status. Ethnic and racial groups have, nevertheless, established some control over the political and economic development processes in many neighborhoods.

DEMOGRAPHIC TRENDS

New York City's population reached its peak of almost 8 million in 1950. After steady population losses throughout the 1970s, as people moved to the suburbs or out of the region altogether, the city's population rose slightly between 1980 and 1990. This increase is attributed both to the stronger local economy of the early to mid-1980s and to the continued attraction of the city to immigrants (Mollenkopf, 1993, p. 97).

White New Yorkers (43%) include Eastern European Jews; Italian, Irish, and Eastern European Catholics; and even a smattering of silk-stocking district Anglo-Saxon Protestants. Although a majority of New York's African Americans (25%) have migrated from the American South, many are immigrants from the Anglo and French Caribbean. Puerto Ricans continue to constitute most of New York's Hispanic community (24%), but their homogeneity is now challenged by a large Dominican population and smaller but still significant groups from Cuba and South and Central America. The Asian population (7%), although still small, has more than doubled over the past decade (Mollenkopf, 1993, p. 97).

CITY FINANCES

New York is not alone in its struggle to maintain a balanced budget. However, its frequent and well-publicized fiscal difficulties reflect, in part, a unique financial situation. New York depends more heavily on federal and especially state aid than does the average municipality, although its revenue patterns are consistent with those of other large cities. The real property tax is a less significant source of revenue in New York than it is in many other municipalities. An array of business and real estate taxes, as well as an unusually (for cities) progressive income tax, supply a significant portion of the City's income. Fees and charges such as licensing fees, sewer charges, and fines contribute an increasingly important share of the City's budget.

New York City spends an unusually high percentage of its budget on services other than those traditionally provided by municipal government, such as police, fire, and sanitation. This is partly a result of the city's historical commitment to support public institutions offering a very broad range of services. Libraries, housing, hospitals, and colleges have historically been funded through city revenues, although the share of city support has generally declined over the past 25 years. Perhaps the most significant obligation on city resources results from the New York State system of financing public assistance, which requires local government to provide one-half of the state's share of welfare and Medicaid benefits. Although New York City has no control over benefit levels, it must foot a significant portion of the bill. The city's social service obligation becomes especially problematic during recessionary periods, when welfare rolls increase and incomes, sales, and business tax collections fall.

THE CITY'S ECONOMY

Like most cities in the Northeast and Midwest, the economic story of New York City over the past generation has been that of an exodus of manufacturing jobs and an employment shift from the goods-producing to the service-producing sector.[1] Unlike many cities, however, New York has always been a center of finance and trade, and it has depended on these trends to retain its economic vitality in the 1980s.

Not only has New York lost shop-floor positions in manufacturing, it has also seen an exodus of the headquarters of many Fortune 500 companies. In 1965, 128 such corporations were headquartered in New York City. By 1988, only 48 remained. Whereas economic and technological changes have encouraged manufacturers and retailers to relocate and disperse their investments, business services—especially banking, legal, accounting, and advertising—have tended to remain within traditional urban commercial centers. Among the largest and most robust employers in New York City are such business service providers. Keeping them in the city has been the focus of the economic development strategy for Manhattan. The current recession, however, has curtailed expansion in the service sector: 32,900 jobs in advertising and data processing; 11,000 in accounting, engineering, and management; and 4,500 positions in the legal profession have been eliminated since 1991. The finance sector has been hit even harder, with 22,400 jobs lost in 1991, although there has been some slow recovery since early 1992. Only health and social services have added any significant number of jobs in recent years.

Government is an important local employer, providing more than 225,000 jobs, mostly held by city residents. In addition, New York State and the federal government maintain regional offices of many of their agencies within the city. However, recessionary pressures and structural imbalances in city finances have led to the loss of more than 36,000 jobs, most of them in local government. Mayor Guiliani slated the elimination of an additional 15,000 jobs for the 1994-1995 fiscal year, a year in which the city's deficit was projected to be in the range of $2.3 billion.

Intergovernmental Relations

STATE AND FEDERAL ROLES

In addition to the strong role in development that economic and commercial forces play, policies are shaped by the nature of intergovernmental relations, as well as by the interests of local officials and community residents. State and federal practices limit the city's capacity to shape its development strategies while also providing, through grants and other kinds of investments, opportunities for certain local actions.

New York State is a major player in city policy making, although it has initiated and sponsored few significant community development programs. By virtue of a home-rule clause embedded in the New York State Constitution since 1924, the State Legislature is allowed to exercise power, particularly veto power, over city governance. In many areas relevant to local community development policy, from setting tax rates to regulating residential rents, state law takes precedence.

The most important state community development agency is the New York State Urban Development Corporation (UDC), which has been involved in several major redevelopment projects in the city. The UDC has state-backed bonding authority and is empowered to bypass local zoning and planning review requirements, making it a useful vehicle for controversial developments. Subsidiaries of the UDC or separate state authorities have been created to oversee major development projects, such as Battery Park City and aspects of redevelopment in Harlem.

Beginning in the 1930s, when Robert Moses's efforts managed to capture one-sixth of the nation's WPA funds, New York City has been adept at securing federal categorical aid for development. The shift to block grants after the passage of the Community Development Act of 1974 altered the federal-city relationship and provided the city with great latitude to employ CDBG funds in new ways. This new discretion allowed the city to dedicate the lion's share of the CDBG budget to maintaining the tax-foreclosed housing stock, which had increased dramatically during the fiscal crisis of the mid-1970s. CDBG funds have not, therefore, been as significant a source of primary revitalization as they have been in other cities (see Rogowsky & Strom, 1989).

Federal CDBG funds have nevertheless been used for other community development purposes. A program in place since the early 1980s uses CDBG monies to revitalize neighborhood commercial strips, improving sidewalks, lighting, and signage in an effort to improve local retail sales. Efforts to retain industry have also been CDBG funded. Services have been improved in traditional manufacturing areas with federal money, and a CDBG-funded revolving loan fund run by the city's development agency has provided low-interest loans for expansion. Smaller federal Economic Development Administration (EDA) programs have also been utilized by city agencies and local not-for-profit agencies to encourage industrial retention.

Urban Development Action Grants have, since their inception in 1978, provided seed money for redevelopment projects. Over a 10-year period, 85 awards were made in New York City; 59 of these have actually been drawn down, for a total of $185.9 million.[2] One-fifth of this total went toward housing subsidy, whereas most of the grants were used to underwrite industrial and commercial development. Generally, the City's Economic Development Corporation (EDC) or its predecessor agencies applied for UDAGs on behalf of project developers and lent the money to the projects at low interest rates. When the loans were repaid, the money was used to fund other redevelopment projects. UDAG-funded economic development projects include the South Street Seaport and the Studio Museum of Harlem, as well as numerous office buildings and industrial projects.

Redevelopment Programs and Their Effects

CITY ECONOMIC DEVELOPMENT
POLICY: STRATEGIES AND TOOLS

The environment in which New York City development policy is created and implemented includes both a complex overlay of public and semipublic agencies and myriad policies and programs. The City Planning Commission, the Department of City Planning, and the city council have major roles in land use decision making. The two city agencies whose primary mission is to foster private investment are the Economic Development Corporation and the Office of Business Services (OBS). With a majority of mayoral-appointed board members, EDC (created in 1991 by merging the former Public Development Corporation, the Financial Services Corporation, and the Department of Ports and Trade) is the city's "deal maker"—the lead agency in any publicly subsidized development project. It is empowered to dispose of city-owned land, lend out city and federal funds, and package other city tax incentives. The OBS (also created by the merger of other units in 1991) focuses on retaining existing businesses by administering subsidized finance programs and technical assistance, including CDBG funds used to revitalize neighborhood commercial strips. EDC, organized as a public benefit corporation, enjoys greater independence from public regulation and oversight; the heads of both EDC and OBS report to the deputy mayor for economic development.

Like most cities, New York lacks major economic levers with which to direct private investment. In addition to acting as a conduit for ever-diminishing federal funds, city officials can offer four kinds of investment incentives. First, they can abate property taxes, and they have done this on most new residential and commercial development— although recently they have tried to restrict abatements to areas with weak property markets. Second, they can abate many of the smaller but significant city business taxes to encourage new investment. Officials can waive the occupancy tax, reduce sales tax on certain items, and offer tax credits to reduce the high local energy costs. Third, they can use city capital funds, either as grants or low-interest loans, to underwrite infrastructure improvements on public land that aid private development. For instance, the city has spent capital money to improve access roads to an office park in Staten Island and to develop a Commons park area for the Metrotech office project in Downtown Brooklyn—one of the case studies presented below. This increases the value of the project to the investor while abiding by the legal restrictions on the use of capital funds. Fourth, tax-exempt bonds are floated by the city itself or authorized agencies; the money earned can be lent to projects at below-market interest rates. In addition to these hard subsidies, city development agencies act as ombudsmen for investors, helping them deal with various city review processes, providing time savings that ultimately reduce private development costs.

During the period studied, city officials used these incentives, as well as zoning changes and marketing campaigns, to convince corporate planners to locate in the four outer boroughs. This strategy gained momentum for other reasons as well. Planners supported the idea of multiple business centers to disperse economic activity and to improve utilization of the public transit infrastructure. Politically, the district-based city council and the borough executives (previously participants in citywide decision making through the Board of Estimate) supported the dispersion policy because it directly benefited their territories. The outer borough strategy also provided Mayor Koch with a means to hold together a governing coalition on the Board of Estimate during the 1980s. It is this convergence of planning and politics that set the stage for the case studies that follow.

DOWNTOWN BROOKLYN AND JAMAICA, QUEENS

Background

The recent histories of Jamaica, Queens, and the section of Brooklyn referred to as Downtown provide instructive examples of the impact of post-World War II economic and social trends on city centers, as well as the capacity of public policy efforts to counteract them.[3] Brooklyn had itself once been a major city boasting its own commercial and civic center. Downtown Brooklyn's significance began to fade after construction of the Brooklyn Bridge (1883) made Manhattan's commercial center accessible and consolidation into Greater New York City (1896) robbed it of its independent civic functions. Nonetheless, the area east of the bridge's Brooklyn terminus has remained important as a borough center. Local banks and utilities retain headquarters in its office buildings, and the retail district along Fulton Street, home until the 1970s to half a dozen national department stores, has long been among the nation's top 10 generators of retail sales. Proximity to Manhattan and accessibility to several subway lines have helped keep the area from falling victim to extensive disinvestment.

Jamaica's central location between Manhattan and suburban Long Island has long been the basis of its regional economic significance. Agricultural and manufactured products en route to Manhattan and Brooklyn from the entire Northeast passed through Jamaica, producing a bustling commercial area. Middle-income residents from surrounding communities, from eastern sections of Brooklyn, and from Long Island viewed Jamaica as their downtown. Macy's, Mays, and the Gertz family's flagship department store were among the retail anchors in Jamaica, and meatpacking and milk processing served to support the downtown's industrial base.

Several related developments led to the deterioration of these areas between the 1950s and the early 1970s. Both centers are surrounded by neighborhoods that have undergone dramatic racial and economic changes. What had been largely white working-class and middle- to upper-class neighborhoods are now predominantly black working-class to poor neighborhoods. Today, the middle class of all races has largely left these areas in favor of the suburbs. The exodus of manufacturing jobs has left the remaining working-class residents economically worse off than those of the previous generation. The demise of numerous local institutions signals the physical and psychological decline of these areas. Both Brooklyn and Queens have lost local newspapers, banks, and, in the former's case, a much-loved baseball team.

"White flight"—the exodus of mostly white middle-class families from Brooklyn and Queens to the suburbs—left a population with far less buying power to shop in these areas. The significance of this population movement is not just economic: The fact that the remaining population was predominantly black fostered a sense among white shoppers that these areas had become unattractive and unsafe. More affluent whites in surrounding neighborhoods—middle-class Queens strongholds such as Jamaica Estates and Kew Gardens and newly gentrified Brooklyn areas such as Park Slope and Cobble Hill—no longer shop downtown in significant numbers. Leaders of revitalization efforts in both neighborhoods have had to counter these prejudices in an effort to encourage whites and even middle-class blacks to venture back.

Almost simultaneously came the growth of suburban retailing. Vast enclosed shopping malls sprouted up on Long Island and in less urbanized sections of Brooklyn and Queens. These malls offered convenient parking, comfortable environments, and a sense of security. Retailers in Jamaica and Downtown Brooklyn could not compete, and department stores began closing their doors in the late 1960s. Changes in the retailing industry unrelated to these local issues forced the closing of other department stores. Replacing large department stores were smaller, more competitive merchants appealing directly to the lower-income groups who were now the principal patrons of these areas. This served to reinforce the reluctance of middle-class shoppers to view Brooklyn's Fulton Street or Jamaica as appealing retail areas.

In the late 1960s, as these trends became more manifest, local businesses, residents, elected officials, and city planners began to explore ways to revitalize Jamaica and Downtown Brooklyn.

Methodology

We chose Downtown Brooklyn and Jamaica, Queens, for study because they have received considerable city, state, and federal investment over the past 25 years and therefore offer an opportunity to assess the impact of public policy on revitalization efforts. In both communities, public and private actors have played significant roles, sometimes separately and sometimes in tandem, providing a view of the partnership aspects of redevelopment. Within each area we studied several projects that spanned the 1970s and 1980s and that enlisted a variety of public and private participants. The projects had multiple interrelated goals, including job creation, support for retailing activity, and the physical upgrade of deteriorating neighborhoods.

We gathered information about each case from several sources. We searched the *New York Times* for relevant articles and gathered historical information from archives of other newspapers. Whenever possible, we reviewed planning documents from city agencies, such as environmental impact statements and urban renewal plans. Finally, we conducted in-depth interviews with identifiable and accessible participants.[4]

CASE STUDY 1: DOWNTOWN BROOKLYN

Context of Revitalization Efforts

Brooklyn's downtown area, served by virtually all of the city subway lines, includes several public and private colleges; a civic center that includes a landmark Borough Hall, the borough's Municipal Building, and state and federal court buildings; several local bank and utility headquarters; and, more recently, the back offices of major financial service firms.

Concerns about the downtown's flagging property values surfaced during the Great Depression, and as early as 1944, urban renewal plans were drawn to accommodate nineteenth-century land use patterns to demands of the twentieth-century downtown. These plans reflect the concerns often seen in postwar renewal efforts. "Blight" was attributed to changing demographic patterns and the elevated subway lines that not only cast a pall over the main streets, but obstructed new development. Large-scale clearance to rid the area of "obsolete" structures (primarily warehouses, factories, and tenement housing) was recommended; indeed, large parts of the plan were carried out. Streets were widened, new public buildings were erected, and middle- and low-income housing developments were constructed. The rationale for these planning efforts, unlike most such efforts today, was not economic develop-

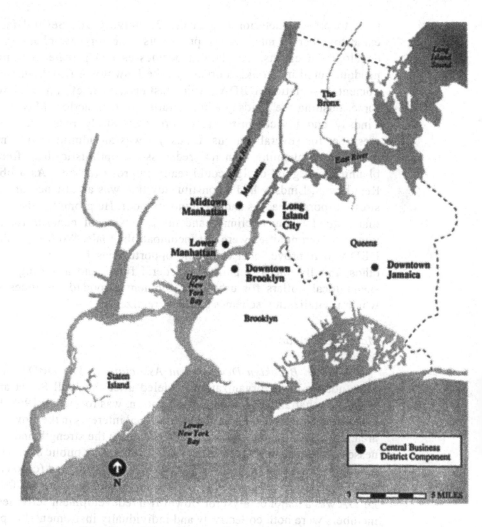

Figure 3.1. Central Business District Components, New York City
SOURCE: New York City Department of City Planning.

ment as much as civic improvement: the "search for a dignified and suitable civic center" for Brooklyn.

By the late 1960s, the continuation of racial transition, white flight, and property disinvestment trends begun in the 1940s gave rise to a new push to prevent deterioration of the Brooklyn CBD. Several factors came together to make this a propitious time for new planning and development efforts. The largest businesses and private institutions headquartered in Brooklyn organized the Downtown Brooklyn Development Association (DBDA), which fast became an effective lobbying force, pushing the Lindsay administration to take action. Mayor John Lindsay and his commissioners were particularly receptive to this message, for several reasons. Lindsay's was an administration more receptive to planning than its predecessors, optimistic that effective planning and good design could really improve the city. As a liberal Republican, Lindsay had a constituency that was an alliance of opposites: corporate leaders and the minority poor. He promoted the notion that a good business climate and the extension of benefits for poor minority communities were not incompatible goals. Working with the DBDA to improve local economic opportunities fit in well with his ethos. Finally, the availability of federal funds and a willingness to spend local dollars for urban development provided resources with which revitalization schemes could be realized.

Major Participants

Downtown Brooklyn Development Association. The DBDA, a private, not-for-profit organization modeled after the Wall Street area's Lower Manhattan Development Association, was formed in 1968. Representing business and nonprofit institutional interests in the Downtown area, the DBDA's purposes were (a) to evaluate the strength and weaknesses of existing resources in the CBD, (b) to lobby public officials for greater attention to the needs of Downtown Brooklyn, and (c) to create a long-term development plan for the area. Into the late 1970s, the DBDA was a major catalyst for Downtown redevelopment activities. Its members were both collectively and individually instrumental in pushing the Lindsay administration to create a special office to oversee Downtown Brooklyn development. Their support was crucial to the adoption of specific projects, such as Fulton Mall (discussed below). When the city's fiscal crisis forced layoffs, the DBDA hired planners who had been working on Downtown projects, enabling them to continue their work without major interruptions.

By the late 1970s, however, the DBDA had begun to lose its effectiveness. The organization became more routinized—whereas CEOs had themselves regularly attended meetings in the late 1960s, a decade later only their staff were attending—and several of the important retail and banking establishments that helped establish the group moved or went out of business. The DBDA's director and board were persuaded to merge with the Chamber of Commerce—by then also an organization of limited effectiveness. As in Jamaica (discussed below), the chamber's concerns were somewhat more focused on helping existing businesses. After the merger, the DBDA became a less active promoter of development activities.

Despite these organizational changes, the ongoing lobbying efforts by Downtown elites were reflected by the Lindsay administration's creation of an agency called the Office of Downtown Brooklyn Development (ODBD) for the Brooklyn CBD. The ODBD was responsible for implementing the plan for Downtown released in 1969, which included office development, housing, transportation, and retail components. When the 1975 fiscal crisis forced a curtailment of virtually all public sector initiatives and a consolidation of the ODBD into a citywide agency, continuity for planning efforts was sustained when some of the Brooklyn-based staff people were transferred to that newly reorganized city development agency, where they continued to push for Downtown projects from a new vantage point. Successive development agencies—most recently the EDC—have maintained their involvement in Downtown Brooklyn.

Public Officials. Brooklyn Borough President Howard Golden and his staff played an essential part in Downtown redevelopment. It is not surprising that the Borough President's Office was an advocate for Downtown. Unlike the other outer boroughs, Brooklyn has one central downtown commercial district, and Borough Hall is located at its heart. Previous borough presidents had been advocates for the borough, but had been more beholden to local concerns and thus less willing to take the chance of participating in more aggressive redevelopment efforts. For example, former Borough President Leone voted against a parking garage that was considered essential to the pedestrianization of Fulton Street (discussed below) when it came before the Board of Estimate, because it involved relocation of one or two merchants. Golden, on the other hand, decided to make his office the catalyst for Downtown redevelopment. His staff played the marketing and promotional role traditionally associated with the borough president, but it did so more aggressively. His vote on the Board of Estimate gave him leverage over budget and land use matters; not only did he have his one vote, but he could broker that vote to get the support of other members on Brooklyn projects. In addition, the support of Tom Cuite, then the powerful majority leader of the city council, allowed the borough president's staff to tuck away dollars for infrastructure improvements and cultural activities that would become the platform on which subsequent development could be built. Finally, the borough president's staff played a crucial brokering role between community groups and developers, preventing the obstruction of some of the more controversial development projects, including Metrotech.

Case Projects

Fulton Mall. Fulton Mall is a pedestrian zone created along Fulton Street, Downtown Brooklyn's traditional retail strip. Approximately 200 stores are located on the strip, ranging from large (Abraham and Strauss's flagship department store) to small (food stores and electronics shops) and including both upscale establishments (the landmark Gage and Tolner restaurant) and bargain basements. Included in the mall area is the enclosed, privately owned Albee Square Mall, developed simultaneously with the rest of Fulton Mall, but under separate auspices.

The Fulton and Albee Square Malls were the city's response to merchants' complaints about the loss of retailing business to malls in the suburbs and outer Brooklyn. If there was a single precipitating event, it was a threat that A&S would move to a location across from Macy's in the new Kings Plaza shopping center at the extreme southern end of the borough. City officials made clear to A&S that the necessary zoning for a second retail center out there would not be forthcoming, but promised that every effort would be made to increase the attractiveness of the downtown area. Pedestrian malls were, in the late 1960s, emerging as the fad in urban retailing; the plan for Downtown redevelopment therefore included the pedestrianization and improvement of Fulton Street as one of its cornerstones. As the plan for the mall evolved, it came to include a two-lane transitway for city buses down the middle, distinctive paving on the widened sidewalks and the road bed, as well as outdoor seating, kiosks, planters, and canopylike bus shelters to give the area a festive look. In October 1973, the proposal for the mall was approved by the DBDA and Fulton Street merchants. The next month, the borough president, outgoing Mayor Lindsay and incoming Mayor Beame, city council majority leader Tom Cuite, and the City Planning Commission all gave approval. A capital budget amendment passed by the city council and the Board of Estimate allocated funds for design. A grant covering 80% of all capital costs was secured from the Federal Urban Mass Transit Administration (UMTA) in 1977, and construction began on the project later that year.[5]

The process for planning Fulton Mall was—by New York City standards, at least—one of overwhelming consensus. Promising to revitalize an existing commercial district, the project entailed no displacement. Key political leaders and private sector influentials had signed on to the idea early in the process. Community groups were either indifferent or, in the case of the local Community Board, which plays an important advisory role on various applications, supportive. The ODBD's goals for the project were undoubtedly shared by all. According to ODBD records, the mall was expected to preserve the Brooklyn CBD and in turn catalyze further office development. Furthermore, it was hoped that this investment would have psychological payoffs. By organizing the Fulton Street merchants and by showing that the city was prepared to make significant investments in the area, this project would make it less likely that local businesses would move away. Because the mall would be constructed largely with federal funds, there were few complaints from those concerned with budgetary matters.

However, the conflict that was absent from the planning phase emerged during implementation. Construction, particularly during the first phase, was, at $24 million, way over budget and behind schedule; when the Christmas season came around, the streets and sidewalks were still torn up and construction equipment blocked many store entrances. Delays resulted from a number of factors, including the fact that engineers had failed to anticipate the complexity of construction problems resulting from the convergence of underground subway and utility lines. The mall was finally finished in 1984.

One of the selling points of the mall for public officials and merchants was that it would be self-supporting. An official merchants association, the Fulton Mall Improvement Association (FMIA), was formed to run the mall, taking care of maintenance, marketing, and security. It was funded through a special tax assessment levied on all property owners in the zone. Based on a similar arrangement in Minneapolis, the concept of a special assessment district was then new in New York State. One of the most difficult tasks of ODBD and DBDA planners was to craft an arrangement that satisfied the State Legislature and various property owners, all of whom naturally hoped to minimize their own liability.[6]

The Albee Square Mall proved even more problematic. Its developer, the Rentar Corporation, had purchased the closed Albee Theater on Fulton Street with the intention of attracting a national department store to the site and building a shopping center around it. Formulated in the late 1960s, this idea fit in with what the city and the DBDA planned for the area. It would provide an anchor for the eastern, more deteriorated end of Fulton Street, where plans were already under way by Con Edison and New York Telephone to build new office buildings as part of an urban renewal plan. The city offered to extend the boundaries of the urban renewal area to include Rentar's mall site.

The intervening years were not good to the plans for the Albee Square Mall. First, no national chain was interested in the site. After several years of searching, Rentar decided to proceed with the mall in the absence of a large anchor tenant and persuaded skeptical city officials that an anchorless mall could be successful.[7] Meanwhile, the fiscal crisis had forced the city to scale back its urban renewal ambitions, eliminating part of the intended mall site. After beginning construction in 1977, the developer had tremendous problems with financing and completed the project in 1980 only with the help of a $16 million EDA loan guarantee and a $2.1 million UDAG. From the start, merchants complained of high rents and poor outreach and maintenance as the mall failed to attract the anticipated consumer traffic. Recently, the original developer sold out to Forest City—developer of Metrotech—whose principals hoped to use their leverage as owners of more prosperous malls across the country to get national tenants to Albee Square.

Today, Fulton Street remains one of the strongest downtown retailing centers in the country. Census figures indicate an increase in retail employment for the Downtown Brooklyn zip code between 1969 and 1987. Reported sales for Downtown Brooklyn place it on the list of the top 10 retailing centers in the nation, and many observers believe these may be underreported.[8] Rents along Fulton Street remain high—between $50 and $75 a square foot—and a visitor would be impressed by the constant crowds at the mall.

It is difficult to know how much of this economic vitality can be attributed to public investment. With few but notable exceptions, the stores on Fulton Street cater to poor and working-class people, primarily black and Hispanic. In spite of efforts to maintain design standards, signage tends toward the garish and the specially designed street furniture has proven difficult to maintain. Many merchants appear to be relatively unconcerned about these amenities, and the mall property owners have, critics complain, been lackluster about upkeep and loath to assess themselves at a rate high enough to allow the FMIA to expand its effectiveness. Yet observers feel certain that, at the very least, A&S would have left the area if the mall had not been constructed, a move that would have triggered a precipitous decline.

Metrotech. The Polytechnic Institute of New York (PINY) and Forest City Ratner Companies, an affiliate of the Cleveland-based Forest City Enterprises, are partners in the 10-building Metrotech development, which has received subsidies from city, state, and federal governments. Planning for what has emerged as a 4.23 million-square-foot, 16-acre development began in the early 1980s, and the first tenant moved in in 1990. The development includes upgraded facilities for PINY as well as office buildings catering largely to computerized back-office operations. As of this writing, six buildings and a courtyard park have been completed and another new building is under construction; the developer is seeking tenants before beginning construction on the remaining sites.

Plans for revitalizing the area around PINY were discussed as early as 1940, and the area was designated for urban renewal in the early 1960s. Implementation was always shelved, however, as attention was devoted to other areas that were either more prominent or needier. Impetus to revitalize the area finally came when PINY officials approached the city and said that they needed to be able to upgrade the Brooklyn campus or they would shift PINY's focus of operations to its suburban Long Island campus. Inspired by the success stories of research and development meccas built around MIT in Cambridge and Stanford in Silicon Valley, PINY officials had the idea of creating a high-tech research center around the downtown Brooklyn campus. Working with DBDA, which received a HUD planning grant, PINY hired a consultant to prepare a feasibility study. Although the city's (then) Public Development Corporation (PDC) was interested in promoting Downtown development, it was skeptical about PINY's capacity to single-handedly pull off a risky development project. PDC refused to begin condemnation and clearance for the urban renewal areas unless PINY could find a developer with which to engage in a joint venture and more concrete plans could be drawn up.

PINY issued a request for proposals, sending packets out to several hundred developers; two responded. One was Bruce Ratner, a former New York City consumer affairs commissioner in the Lindsay administration and part of the family that owned the well-established Forest

City Enterprises. Forest City, whose first home base was ironically located in the Brooklyn section of Cleveland, had made its reputation building shopping malls throughout the Midwest. Working through its New York cousin, it was interested in gaining a toehold in the booming city real estate market and gambled that Brooklyn would present such an opportunity. Based on its years of development experience, commitment to stay in the area, and deep pockets, PINY and the city's PDC chose Forest City.

The evolution of the Metrotech idea from research and development facility to office campus was the result of Forest City's marketing sense, real estate trends, and city policy. In the early to mid-1980s, the Manhattan real estate market felt the effects of the exploding financial services industry. Rapidly increasing rents were making it difficult for companies to expand or renew leases within Manhattan. The city's outer borough strategy centered on ways to encourage companies to keep their land and labor-intensive office operations in New York City. Staff time and financial incentives were devoted to identifying and packaging development sites outside the traditional central business district. Forest City officials saw the possibility of taking advantage of these existing programs and tailoring the project to just such computer-intensive, back-office users. This allowed them to market what they felt were Metrotech's locational advantages. Its proximity to Lower Manhattan, emphasized in the "Wall Street East" promotional campaign, made it easily accessible to headquarters and other ancillary services. The convenience to several subway lines made it an easy commute for office workers more likely to be living in Brooklyn and Queens than in Manhattan. The association with PINY, whose computer science programs were well respected, was an additional selling point for technology-minded managers. Finally, Forest City used design as a marketing tool, offering buildings that accommodate modern communications infrastructure and investing significantly in the creation of public areas attractive enough to transform outsiders' perceptions of the area.

City policies offered considerable aid to the Forest City-PINY marketing efforts. Subsidies offset early development costs, and further subsidies were available directly to tenants. Public development subsidies included two UDAGs totaling $14 million, $30 million from the city's capital budget (to be repaid by the developer according to a fixed schedule), $3 million in low-interest industrial revenue bonds, a $14 million loan from the Port Authority of New York/New Jersey toward the cost of PINY's library, and a $10 million loan from the Municipal

Assistance Corporation. In addition, operating subsidies for the developer and the tenants included waiver of the city's occupancy tax for 12 years, a $500 annual tax credit per employee, real estate tax abatements phased out over 23 years, energy cost reductions, and discretionary sales tax abatement. Whereas some of these items represent as-of-right city programs available to all new development, others resulted from negotiations involving Forest City, PINY, and the city's PDC. On this basis, Forest City has successfully signed up major tenants, including the Security Industries Automation Corporation (which processes stock market transactions for the New York and American Stock Exchanges), Brooklyn Union Gas, and New York Telephone.

In 1989, Chase Manhattan announced its decision to move most of its operations out of New York City. Koch administration officials began frantic negotiations to keep Chase's 6,000 jobs, offering a package of subsidies totaling more than $200 million over 15 years, and arranging for Chase's move to two previously unclaimed Metrotech sites. The arrival of Chase virtually assured the success of the project; additional tenants have been signed up since that deal was finalized.

Unlike the Fulton Mall, Metrotech is a massive project that required the closing off of streets, rerouting of buses, displacement of small businesses and residents, and commitment of millions of public dollars. The planning process for the project was therefore contentious. Metrotech's ultimate approval involved negotiations among the city, the developer, and various local constituents, with the Borough President's Office frequently acting as mediator.

The Borough President's Office and Forest City together managed to create a supportive coalition among those elements of the community not directly in opposition.[9] The Downtown business community, no longer as dominant a force as it had been in the late 1960s, remained supportive. Forest City took the lead in organizing a Business Improvement District, including the Metrotech development and a few surrounding institutions, in order to provide a high level of services and organizational support for a new wave of advocacy for Downtown Brooklyn business interests. The leaders from the adjoining working-class black communities were invited to address the concerns of nearby residents who wanted to gain from economic benefits growing out of Metrotech. Forest City established an affirmative action program to facilitate access to construction jobs and hired staff to help neighborhood residents find out about and prepare for employment opportunities. What is perhaps most important from a planning viewpoint is that the efforts of the developer—facilitated by the Borough President's Office and the Community Board—turned a constituency that could have created obstacles to the project into active supporters or, at the least, passive bystanders.

The size of the Metrotech project means that its impact on Downtown Brooklyn will ultimately be enormous; even today, increased subway ridership and pedestrian traffic are noticeable, and perceptions of the area have begun to change. Ultimately, the area's retailing should respond to the presence of thousands of additional employees, which is one reason Metrotech's developers have become active in trying to improve the quality of retailing in Fulton and Albee Square Malls.

There is considerable public subsidy in the Metrotech project, totaling $71 million. Virtually all of this money is structured as some kind of loan, so, barring an unforeseen reversal in the project's fortunes, it will be paid back. Some 15,000 employees are expected to work on-site, with an estimated annual payroll of $234 million, and an equal number of jobs may be generated off-site. Based on these assumptions, the following financial benefits to the city have been projected: $5.7 million in income taxes during construction and $5.7 million annually from ground lease payments. Assuming that 87% of the employees are city residents, local income tax revenues will ultimately come to $3.7 million a year, and proceeds from the commercial occupancy tax will add an additional $3.5 million a year. Real estate taxes, abated for 10 years and then phased in gradually, will, in year 13, yield $1 million in city property taxes and $40 million (in 1985 dollars) by year 27. Largely because of the unanticipated bonanza of the Chase relocation, Metrotech is, as of now, ahead of schedule. The current glut of office space in lower Manhattan makes it unlikely, however, that there will be large tenants shopping for space in the near future.

Metrotech's impact has already been felt on the political landscape of Downtown Brooklyn. Forest City Ratner is now the major player in development politics. Since its designation as the Metrotech developer, Forest City has completed another large office tower (at the edge of Brooklyn Heights and the civic center) for a financial services firm, bought the Albee Square Mall, was named developer for and has completed the renovation of a federal building, and has entered a joint venture with another developer on a major mixed-use project a few blocks from Metrotech that had been stalled for more than a decade. Its influence has increased as the DBDA has receded. Even though the borough president's power has waned with the abolition of the Board of Estimate, Forest City officials hope that the presence of the influential corporations that are the major tenants of the new developments will give added clout to lobbying efforts aimed at improving local services.

CASE STUDY 2: JAMAICA, QUEENS

Context of Revitalization Efforts

Jamaica rests at the geographic heart of Queens, located just minutes from several major arteries to and from eastern Long Island and the sister boroughs. Four subway lines, 42 surface routes, and 10 of the 11 Long Island railroad paths stop in Jamaica—attesting to its importance as a transportation hub. It is conveniently located near LaGuardia and Kennedy Airports. Jamaica's public facilities include York College of the City University of New York, Queensborough Central Library, three major hospitals, and several state and federal agencies—most notably, the recently completed Social Security Administration's regional headquarters.

The initial phase of Jamaica's redevelopment began in 1967 with efforts from public and private sector actors to use public investment as a catalyst for a broader revitalization effort. Representatives from government and the private Fund for the City of New York and the Jamaica Chamber of Commerce were instrumental in convincing the influential Regional Plan Association to conduct a study of the economic potential of Jamaica, an important first step in creating a development plan. The study encouraged the redevelopment of Jamaica as a commercial center, building on its transportation infrastructure and central location between Manhattan and Long Island. It endorsed the promotion of a newly formed public-private partnership, the Greater Jamaica Development Corporation (GJDC), as the lead agency to develop and market the area.

Three major public improvements, deemed the heart of efforts to attract private capital, were pushed by Jamaica advocates beginning in the late 1960s. All were finally completed by the mid-1980s. Planners claimed that the elevated subway line running over Jamaica Avenue discouraged commercial traffic and, therefore, investment. In 1973, the city appropriated funds for its removal. A new underground line was to be built as a replacement, funded with a $350 million grant from the federal Urban Mass Transit Administration (UMTA) and matching funds from the regional Metropolitan Transit Authority (MTA). The work was finally completed in 1988. Second, Queens officials fought to make Jamaica the home of York College. City University officials fought against this plan, at times claiming that a new campus was unnecessary and at other times fighting for a more suburban location. But local political pressure prevailed. The college's science complex was completed in 1974 and the remainder of the campus in 1986. Finally, the federal government agreed to build its regional headquarters for the Social Security Administration, a 900,000-square-foot, $110 million complex employing 3,500, in Jamaica.

Major Participants

Greater Jamaica Development Corporation. Planning and development efforts in Jamaica have virtually all been initiated and coordinated by the Greater Jamaica Development Corporation, a private, not-for-profit agency that frequently works with or on behalf of city, state, and federal agencies. Its initial strategy, calling for public investment to set the stage for further development that would take advantage of the area's skilled labor force and strategic location, followed the design in the Regional Plan Association's 1968 report.

The GJDC's presence over a quarter century has provided the steady organizational infrastructure on which community revitalization efforts could grow. Relying on public funds, private contributions, and fees for services, the GJDC has sponsored dozens of redevelopment projects focusing on job creation—a goal attractive to elected officials, city agencies, and local residents. One important milestone in its efforts was the area's designation by New York State in 1987 as an economic development zone and the GJDC's appointment as the zone's primary administrator. Businesses within this zone were granted a number of incentives, including abatement of New York State taxes and utility rate reductions.

The GJDC's work has generally received support from elected officials and community leaders. There have been, however, periodic tensions between the GJDC and the Jamaica Chamber of Commerce, reflecting the occasionally divergent interests of existing local businesses represented by the chamber and new development promoted by the GJDC. (Similar conflict was encountered in Brooklyn, between the Downtown Brooklyn Development Association and the Chamber of Commerce.) For example, the Chamber of Commerce had hoped to use economic development zone incentives to help existing businesses improve their properties and upgrade workers' skills. In contrast, the GJDC believed the incentives were better packaged to market the area to new private investment.

Public Officials. The Queens Borough President's Office has been an important actor in Jamaica's revitalization efforts. Borough presidents had, prior to the 1989 charter revision, been well placed to advocate on behalf of a borough business center; they could be both cheerleaders and deal makers, bringing the public sector together with private and community interests that shared a stake in a vibrant downtown. Donald Manes, an active promoter of Jamaica development before his well-publicized death in 1986, was a prime supporter of the GJDC. He ensured the placement of the York College campus in Jamaica and lobbied federal officials to bring the Social Security Administration to the area.

Citywide officials in both city and state agencies have also played important roles in Jamaica, although local informants have noted that their area has not always received the kind of attention or dollars committed to other areas. The city's EDC was instrumental in creating the College Gate, an underpass from the entrance of York College to Jamaica Avenue, designed to attract nonresident students to the downtown center. The state economic development zone designation and the siting of several large agencies in the former Gertz Department Store have both been integral parts of the overall Jamaica redevelopment strategy.

Jamaica's congressman, the late Joseph Addabbo, lobbied strenuously on behalf of revitalization efforts, especially the federally funded Archer Avenue subway and the Social Security Administration building. A federal NEA grant provided for public sculpture in the downtown center and CDBG funds contributed to physical improvements along Jamaica Avenue. The Economic Development Administration (EDA), a partner in the Farmers' Market (see case study below) appropriated funds for the construction of a parking garage for use by shoppers and Social Security Administration employees. Federal impact has been limited by cutbacks in development programs; in some cases, the city's own fiscal problems have held up federal grants that required local matching funds. Ironically, Jamaica's success in preventing dramatic deterioration has proven an obstacle to the area's winning public funds: It is not devastated enough to serve as a symbol of urban decay and therefore is not a prime recipient of city, state, or federal aid.

Case Projects

Gertz Plaza Mall (Jamaica Plaza Mall). The Gertz Department Store served generations of middle-income shoppers in the heart of Jamaica; today the site houses state and city agencies in a 400,000-square-foot office complex built atop two floors of retail space. The Gertz Department Store closed its doors in 1980, at a time when public improvements, including the new subway line and federal building, were promised but not yet under way. Unable to compete with large malls for middle-class shoppers, the Gertz family chose not to renew the store's lease. Coming just three years after the Jamaica Macy's closed, Gertz's decision led to panic among Jamaica's merchants. Alexander's Department Store came close to leasing the building, but at the last minute decided against Jamaica.

The building stood vacant for several years as plans for its reuse were considered. Renovation plans were ultimately made possible by the commitment of public agencies to lease space. New York State had been trying to disperse its downtown agencies out of the World Trade Center in Lower Manhattan since the 1970s. Local elected officials, especially Borough President Manes, convinced Governor Cuomo—himself a Queens native—and Mayor Koch to commit city and state agencies to lease space in Jamaica. The owners hired a developer who, with $25 million in private financing, embarked on a gut rehabilitation of the old building, creating offices above a two-story retail mall.

The borough president and the GJDC sought public agency offices to bring workers and their buying power to Jamaica. The hope for the retail mall was that it would give the new workforce located at the complex a place to shop and, even more ambitious, lure back middle-class consumers from surrounding areas. The GJDC, responsible for marketing the retail space, purposely set rents low—$12 per square foot, half of the typical rent for Long Island malls—to fill up the space. So far, the retail mall has received mixed reviews. Space has largely been rented by both local merchants and larger regional chains, including Fayva Shoes and Sterling Optical. And the shoppers it draws are a lively addition to the streets of Jamaica. Hopes of attracting the kinds of national chains that would draw weekend and evening shoppers, however, have not been realized.

The Farmers' Market. The most innovative of the revitalization projects in Jamaica is the so-called Farmers' Market, which combines a produce market occupied by upstate farmers and a food court for gourmet and fast-food franchises. Its current site has been developed through a $7 million public-private cooperative effort aimed at generating additional downtown activity.

The Farmers' Market has operated at three different locations since its inception. The GJDC conceived of the market as a scheme to build on Jamaica's traditional role as a food-processing center and trade hub. In 1976, the GJDC initiated the effort by inviting several regional farmers to sell their produce on a cleared urban renewal site (later to be designated for the Federal Office Building). The farmers canvassed were initially reluctant to come to Jamaica, which they associated with high crime rates; and at that time the notion of an urban farmers' market was new to the area (although today such markets exist throughout New York City). Moreover, the success of similar efforts in midwestern cities such as Chicago was insufficient to encourage the farmers to set up shop in Jamaica. A few took the chance, however, and their success inspired the GJDC to continue the effort.

In 1981, the original site of the Farmers' Market was appropriated for the construction of the Federal Building, a cornerstone of the revitalization strategy, and it was agreed that the market would be relocated to an adjacent cleared site. At this time, the GJDC received funding from Banker's Trust to conduct a feasibility study about possible expansion of the market and its potential impact on the area's revitalization. The study demonstrated that Jamaica, providing its office and commercial redevelopment hopes were realized, would be a gainful site for a permanent produce market, as well as for fast-food restaurants.

Participants envisioned a permanent market that would link the disparate redevelopment projects throughout the area. With new offices planned in Gertz Plaza, the Federal Building, and Site 1, additional lunchtime facilities would be needed.[10] Early proponents of the market also hoped to attract residents and employees from surrounding neighborhoods to a year-round market in an increasingly active downtown. Finally, additional jobs would be created in the expanded, permanent market, particularly in fast-food establishments.

Support for the project came from many sources. To acquire the privately owned site, the GJDC received grants from two banks. The borough president and mayor pushed through a $1.9 million city capital budget appropriation for infrastructure renovations, and the New York State Urban Development Corporation provided $1.1 million during the early phases of construction. The federal EDA, already invested in the neighborhood through its financing of an adjacent parking garage and vigorously lobbied by Congressman Addabbo, contributed $1.36 million to facilitate the involvement of minority and women-owned businesses, and an insurance company added an additional $500,000 for the same purpose. Largely because of the GJDC's outreach efforts, all of the current fast-food vendors in the market are minority- or women-owned businesses.

The permanent market has been in operation since 1991. It is run as a joint venture between the GJDC and the city's Economic Development Corporation. The GJDC owns half of the market outright and leases the other half from the EDC. In addition, the EDC receives 50% of the GJDC's net income from the rents paid by the farmers. It is estimated that job creation is running ahead of projections; although the food court is not fully rented, the goal of providing a social mixer has been realized.

Policy Implications

ASSESSMENT

Our case studies indicate that urban revitalization efforts involve a variety of local factors interacting with market forces in a complex pattern of planning, decision making, and implementation. Both in Downtown Brooklyn and Jamaica, Queens, a number of separate projects, carried out over long periods of time, have come together to result in large-scale area redevelopment. Yet, with some differences, neither Downtown Brooklyn nor Jamaica has been totally rebuilt or renewed, suggesting the daunting magnitude of revitalization needs in large cities.

The removal of blight, the halt of decades of deterioration and decline, and the reversal of the downward spiral of disinvestment are themselves significant accomplishments. These efforts, which also generate new positive social and psychological perceptions, require extraordinary resources and sustained local political leadership and advocacy. In our case study areas, community-based organizations provided the continuity of interest and commitment to see the projects through. The participation of business leaders and organizations, in funding specific tasks and in creating public-private partnerships, was also vital.

Propitious market forces aided the success of Metrotech. Close proximity to Manhattan, the presence of good transportation routes, and access to labor are factors not easily changed by public policies or programs. But the actions of local participants influenced outcomes as well. Brooklyn and Jamaica both benefited from the support of key elected officials and organized private sector stakeholders. The GJDC has provided a quarter of a century of leadership in the Queens redevelopment area, and the DBDA gave early voice to need for Downtown Brooklyn rehabilitation. Political advocacy, by borough presidents and an influential congressman, proved especially crucial because significant public investments were deemed essential to encourage private development. The ability to leverage city, state, and federal funding, whether in the form of subsidies to private developers or, as has been more common in Jamaica, with new public facilities, has determined whether large-scale revitalization was possible.

ROLE OF FEDERAL PROGRAMS

Federal resources played an important role in both communities, but were a more important factor in Jamaica. Both Brooklyn projects received federal funding: 80% of Fulton Mall's first phase was paid for by UMTA, and a $2.1 million UDAG facilitated the completion of the Albee Square Mall. Two Metrotech buildings received combined UDAGs amounting to $14 million. Clearly, federal spending was not unimportant to the Brooklyn projects. Federal decisions, however, figured more prominently in the Jamaica case. Indeed, Jamaica's progress has been far more tied to public investment from all levels of government. The $350 million UMTA grant for the Archer Avenue subway dwarfs all other federal investments discussed here. CDBG funds paid for commercial revitalization along Jamaica Avenue and helped to support the GJDC. The EDA helped to fund a much-needed garage and is one of the sources of support for the Farmers' Market. But the most significant federal action has been the construction of the Social Security Administration office. Jamaica's advocates still hope this investment will be the catalyst for further private interest in the area.

RECOMMENDATIONS:
INTERGOVERNMENTAL COMMITMENT TO CENTRAL CITIES

The magnitude of federal funding in both case study areas strongly suggests that large-scale urban redevelopment efforts cannot be considered or sustained without the availability of these funds, in comprehensive coordination with state and local efforts. We therefore recommend the following:

• The federal government needs to make a specific and sustained funding commitment to central-city revitalization. The needs of older cities, especially in the Northeast and mid-Atlantic regions, are growing well beyond the ability of municipal governments to raise revenues and deliver basic services. Aging infrastructures requiring extensive repair and, in many cases, the impacts of immigration add to the responsibilities of local government. Large-scale redevelopment of central-city areas must be seen as a problem requiring the attention of larger governmental jurisdictions. Only the federal government can marshal the resources appropriate to the magnitude of these problems. The well-being of older central cities is a national problem. As vital centers of finance, trade and distribution, and culture, these cities are important cogs in their regional economies, affecting the growth and well-being of the suburbs that surround them, as well as essential subnational centers in the global economic network.

• In addition to a significant increase in the level of funding, federal programs need to be targeted to the unique circumstances of older central cities. The CDBG program, for example, should have a separate set of criteria appropriate to these needs. The UDAG program, or a new similar effort, should be a means to enable public-private partnerships.

• State governments have attempted to fill, however partially, the gap created by the decline of federal interest in and support for central cities. The difficulty of raising taxes and the politics of urban versus suburban areas circumscribe the ability of state governments to act. Yet older central cities are key to their states' economic well-being. States should form alliances to advocate for federal programs and to advance the prospect of regional cooperation for comprehensive economic development planning and implementation.

• Community-based organizations, especially local development corporations, have proven to be important, effective, and necessary neighborhood-level actors in the urban revitalization process. For those groups with proven track records, a modest funding stream combining federal, state, and city resources should be in place to guarantee sufficient administrative capacity for these organizations to continue and expand their efforts.

Conclusion:
Case Studies in Theoretical Perspective

Analysis of the planning process in our case studies reflects interesting variations on the themes found in the urban development literature. Although these cases do not disprove the contention that market forces and economic elites exert heavy influence on local development outcomes, they do suggest that the relationships among private investors, public officials, and other participants are more complex. In comparison with other U.S. cities, New York has a fragmented business elite and a large public sector. Elected officials are adept political entrepreneurs; agencies are professionally staffed, and public authority can channel the course of private investment. The planning process therefore is far more driven by public actors in New York than it is in cities such as Atlanta (Stone, 1989) or Dallas (Elkin, 1987). This is especially true for development in the outer boroughs. Except for banks and utilities that are unable to relocate, private investors may not care whether their businesses are located in one place or another, as long as they can be viewed as "profitable." Public officials, however, do care about location and the symbolic appearance of physical well-being. The motivation for redeveloping Downtown Brooklyn and Jamaica, Queens, thus grows out of political or public policy needs, not market forces, and it is therefore not surprising that the case studies presented here feature city agency staff, elected officials, and community-based organization leaders as key initiators of redevelopment efforts. The precise contours of development policies are determined by the political dynamics within particular cities. The availability of federal funding programs, along with the initiatives of local officials, are then the key mechanisms that trigger, enable, and sustain urban revitalization efforts.

Notes

1. Material on the city's economy is drawn from Winokur (1992).
2. The other 26 were canceled or recaptured by HUD for a variety of reasons.
3. The case study narratives were developed based on extensive interviews with local participants and press searches. See note 4 for a complete list of the interviewees. Local press sources are not cited individually in text, but are included in the reference section.
4. To protect the confidentiality of our sources, we do not attribute specific statements to any particular individuals. Regarding Downtown Brooklyn, we conducted in-depth interviews with the following persons: Hardy Adasko, assistant vice president of the New York City Economic Development Corporation (June 9, 1992); Joan Bartolemeo, executive director of the Brooklyn Economic Development Corporation (June 16, 1992); Jill Kelly, executive director of the Fulton Mall Improvement Association (May 22, 1992); Gayla Merriman, project manager, Metrotech (June 25, 1992); Bob Ohlerking, former director of the Downtown Brooklyn Development Association (June 17, 1992); Richard Rosan of the Urban Land Institute, Washington, D.C. (July 2, 1992); Harvey Schultz, former executive assistant to Brooklyn Borough President Howard Golden (July 9, 1992); Michael Strasser, assistant commissioner of the New York City Department of

Transportation (July 6, 1992); Paul Travis, chief operating officer of Forest City Ratner (July 17, 1992); Evelyn Williams, district manager of Community Board 2 (July 7, 1992); Bernd Zimmerman of the Office of Bronx Borough President (June 24, 1992); and Marlene Zurack of the Economic Policy and Marketing Group, Office of the Deputy Mayor for Finance and Economic Development, New York City (July 23, 1992). Regarding Jamaica, Queens, we interviewed the following people: Douglas Brooks, director of the Brooklyn Office of the Department of City Planning (July 10, 1992); Joyce Coward, director of the Technical Assistance Program, New York City Department of Business Services (July 9, 1992); Peter Engelbrecht, director of design and construction with the Greater Jamaica Development Corporation (July 2, 1992); Helen Levine, executive vice president of the Greater Jamaica Development Corporation (June 8, 1992); Peter Magnani, deputy borough president under Queens Borough President Claire Shulman (July 6, 1992); Carlisle Towery, executive director of the Greater Jamaica Development Corporation (June 25, 1992); and Bridget Wieghart, assistant vice president for Queens development, New York City Economic Development Corporation (July 9, 1992).

5. UMTA funding was possible because of the retention of a bus transitway. Those interviewed about the project differed in their recollections of how the funding influenced the planning process. Whereas some recall that the transitway was included to make the project eligible for UMTA funding, others claim that it was necessary in order to keep bus traffic moving and would have been included regardless of the funding source.

6. Such districts have since been institutionalized through state legislation, so that new areas, now called business improvement districts (renamed, no doubt, to provide an acronym—BID—less lugubrious than that for special assessment district, or SAD) can be formed without such extensive negotiations.

7. The developer claimed that centers filled only with small retailers were popular in Europe and were the wave of the future in the United States. Indeed, in 1979, Boston's Faneuil Hall opened without a large anchor tenant, becoming the first of several successful "festival marketplaces" in U.S. cities. Albee Square has never, however, achieved similar success.

8. Retail figures are based on the information merchants report for tax purposes. Several knowledgeable informants are convinced that some of the newer independent Fulton Street merchants may not be fastidious about reporting sales to state officials.

9. The only sustained effort against the Metrotech project came from a small but very vocal group of artists who lived in converted industrial lofts that were scheduled for demolition. They, along with a few local businesses also designated for relocation, waged a battle in the courts based on clean air/environmental impacts. In a predawn meeting with city officials, tenants, and the developer, Borough President Golden facilitated an agreement with the artists, other tenants, and local businesses for generous relocation benefits and the rebuilding of a local church in the area.

10. Cleared as part of an earlier urban renewal effort, Site 1 is a vacant lot adjacent to the Federal Building. Plans for the construction of a publicly financed, privately owned office building targeting back-office uses fell apart in the late stages of development. Construction of the building was predicated on assurances from the city about completion of the new subway station and the Federal Building, and the developer was unable to convince city officials to extend his timetable for beginning construction when these other projects were continually delayed. As a result, he lost his private financing commitments, and the project was abandoned.

References

Carrying the torch for downtown growth. (1991, October 27). *New York Times.*
City agency opens bidding for prime Jamaica location. (1987, July 17). *New York Newsday.*
Coming attractions for office tenants. (1987, June 28). *New York Times.*
Developing a new style for downtown Jamaica. (1985, February 27). *New York Newsday.*
Elkin, S. (1987). *City and regime in the American republic.* Chicago: University of Chicago Press.
High tech Bklyn complex approved. (1987, July 2). *New York Times.*
Hoot at Jamaica "ghost town" charge. (1984, February 24). *New York Daily News.*
Hopeful signs of upturn lifting spirits in Jamaica. (1985, July 18). *New York Times.*
Jamaica, a success story still waiting to happen. (1980, July 6). *New York Times.*
Jamaica Ave. lives, Gertz or not. (1980, April 14). *New York Daily News.*
Jamaica is the toast of the town. (1988, June 12). *New York Daily News.*
Jamaica project collapses. (1987, February 2). *New York Newsday.*
Jamaica's Farmer's Market moves; makes way for office building. (1986, May 17). *Forum.*
Mayor makes pitch for city's outer space. (1983, September 24). *New York Newsday.*
Metrotech wants approval amid protests. (1987, August 15). *New York Times.*
Mollenkopf, J. H. (1993). *New York City in the 1980s: A social, economic, and political atlas.* New York: Simon & Schuster.
New York City Planning Commission. (1993, January 6). *Shaping the city's future: New York City planning and zoning report* (draft). New York: Author.
Office/retailing complex to rise in Jamaica. (1985, October/November). *New York Page.*
Once again, the search for a developer. (1987, April 26). *New York Times.*
Planning strategies for a new retail environment. (1992, June 14). *New York Times.*
A private sector boost. (1985, August 11). *New York Times.*
Queens renaissance: Gertz store going mall. (1985, September 29). *New York Daily News.*
Rebuilding downtown Jamaica. (1985, December 24). *New York Times.*
Rebuilding Jamaica. (1984, June 21). *Queens Ledger.*
Rogowsky, E. T., & Strom, E. (1989). *Improving CDBG services through university involvement: A study of the community development process in New York City.* New York: City University of New York, Graduate School and University Center, Robert F. Wagner, Sr., Institute of Urban Public Policy.
Still spring in farmers' Jamaica. (1989, November 19). *New York Newsday.*
Stone, C. (1989). *Regime politics: Governing Atlanta, 1946-1988.* Lawrence: University Press of Kansas.
30MM rehabilitation of the former Gertz department store creates new office and retail complex in Jamaica, New York. (1987, October/November). *New York Page.*
3 major projects set for Jamaica. (1983, June 16-22). *Queens Tribune.*
Transforming downtown Brooklyn. (1989, January 22). *New York Times.*
2 acres in Jamaica. (1987, July 12). *New York Times.*
U.S. says most of growth in 80's was in major metropolitan areas. (1991, February 21). *New York Times.*
Wanted: New developer. Shulman seeks bidders for delayed office building. (1987, February 13). *New York Daily News.*
Will bitter rivalry jeopardize Jamaica development zone? (1988, March 21). *New York Newsday.*
Winokur, L. A. (1992, October). Who's working where in New York? *Empire State Report,* pp. 26-31.

4

The Revitalization of New Orleans

MICKEY LAURIA
ROBERT K. WHELAN
ALMA H. YOUNG

New Orleans was a late starter in federal urban revitalization efforts. Prior to the Landrieu mayoral administration in 1970, state and city officials viewed federal dollars with suspicion. They did not want the federal government meddling in state and local affairs, and they were specifically concerned with what an influx of federal dollars would do to the politics of maintaining the state and local governing coalition. The main focus of this chapter will be an analysis of the use of federal funds in New Orleans after 1970. Our review will center on the mayoralties of Maurice "Moon" Landrieu (1970-1978), the late Ernest "Dutch" Morial (1978-1986), and Sidney Barthelemy (1986-1994).

Mayor Landrieu realized that he could use federal funds to bring African Americans into the city's government and politics, without antagonizing conservative white voters. Landrieu primarily used federal antipoverty funds and funneled them through black political organizations (Whelan, Young, & Lauria, 1994, p. 5). It was not until the Morial administration in 1978 that the use of federal funds for economic development became a priority (Whelan et al., 1994, p. 6). It was this shift, the city under Morial's leadership assuming a major role in economic development, that began to create problems within the governing coalition from conservative white components. Morial's white support subsequently dwindled. He was able to maintain the mayorship because of the black electoral coalition he had consolidated. With the withdrawal of federal dollars in general in the 1980s and the increased emphasis on private sector leverage of federal economic development dollars, Barthelemy returned the development leadership to the private sector in general and, specifically, encouraged the growth and development of quasi-public development organizations (Whelan et al., 1994). Thus the nature of the governing coalition shifted to a corporate-centered arrangement (Brooks & Young, 1993). Consequently, although federal programs have been instrumental in revitalizing urban areas throughout the United States, their role in New Orleans has been less consequential.

City Context

HISTORICAL BACKGROUND AND POPULATION

New Orleans is a city whose place in the urban hierarchy has changed over time. Its original growth was based on its port near the mouth of the Mississippi River. New Orleans was once one of the nation's leading cities; in 1840, for example, it was the third-largest city in the country. In the 1990 census, New Orleans, with a population of 497,000, had the 24th-largest central city in the country. New Orleans's place in the region has changed considerably, too. Until the 1950 census, when it was passed by Houston, New Orleans was the largest city in the South. In the 1990 census, the Texas cities of Dallas, El Paso, Houston, and San Antonio were found to be larger than New Orleans. Jacksonville, Florida, and Memphis, Tennessee, also have bigger central-city populations. Perhaps the New Orleans metropolitan area is more reflective of the regional situation. The New Orleans metropolitan statistical area, with a population of 1,285,000, is the 31st-largest overall. It is behind the southern metropolises of Atlanta; Dallas-Fort Worth; Houston; Miami-Fort Lauderdale, Florida; Norfolk-Virginia Beach-Newport News, Virginia; San Antonio; and Tampa-St. Petersburg-Clearwater, Florida. The central city gained population between 1950 and 1960, but since 1960

TABLE 4.1 Population, Orleans Parish

Year	Population	Population Change		
		Year	Number	%
1950	570,445			
1960	627,525	1950-1960	+57,080	+10
1970	593,471	1960-1970	-34,054	-5
1980	557,927	1970-1980	-35,544	-6
1990	496,938	1980-1990	-60,989	-11
Total		1950-1990	-73,507	-13

SOURCE: U.S. Department of Commerce (1993, Table 30).

it has been losing population. Its most dramatic losses have taken place since 1980 (see Table 4.1).

New Orleans has always been a city of great ethnic diversity (see Hirsch & Logsdon, 1992). However, the two major ethnic groups remain African Americans and whites. In the years between 1960 and 1990, the racial composition of the city reversed itself. In 1960, whites were 62.6% of the city's population, and blacks were 37.2%. As of the 1990 census, blacks were 61.4% of the city's population, and whites were 33.1% (Mumphrey & Akundi, 1994). There also are growing numbers of Hispanics and Vietnamese in New Orleans.

At the same time this racial shift has been taking place, Orleans Parish has undergone what Harrison and Bluestone (1988) have termed the "Great U-Turn" (see also Smith & Keller, 1986). The parish's economy has polarized, with more people living in affluence and greater numbers of people living below the poverty level. In 1960, the proportion of families with annual incomes of less than $3,000 living in Orleans Parish was 24.2%. By 1970 the proportion, now standardized as being below the poverty level, had decreased to 21.6%. It remained stabilized through the 1980s at 21.8%, but by 1990 it had increased dramatically, to 31.6% of Orleans Parish families living below the poverty level.

LOCAL GOVERNMENT STRUCTURE
AND INTERGOVERNMENTAL RELATIONS

New Orleans is a "home-rule" city with a strong mayor form of government. The mayor has the power to appoint department heads, authorize contracts, initiate the budget process, and veto council actions. It takes a vote of five out of seven council members to override a mayoral veto. The mayor is elected for a four-year term and is limited to two consecutive terms. The seven members of the city council (five district and two at-large) are also elected for four-year terms. For a city the size of New Orleans, the council is relatively small. Those holding political office are becoming more reflective of the city's racial makeup. The last three mayors have been African American males. The 1990-1994 city council comprised three African American males, one African American female, one white male, and two white females. A term limitations law resulted in considerable change in council experience, as only two incumbents kept their seats after the 1994 council election. The new 1994 council consists of four African American males, one African American female, and two white females. The two white women are Republicans; the other council members are all Democrats.

The governments of the city of New Orleans and Orleans Parish are consolidated. This city-parish consolidation is often ignored by students of local government, perhaps because it has been in place since 1805. The major areas of overlapping jurisdiction are in law enforcement and judicial areas, with the city's police department complemented by the parish's civil and criminal sheriff's departments.

New Orleans's relationships with other parishes are problematic, on all sides. Former Mayor Morial pursued an openly conflictual approach with suburban jurisdictions. For instance, he proposed an income tax on salary earnings in New Orleans, arguing that those who live in the suburbs but make their livelihoods in New Orleans should pay their fair share. Mayor Barthelemy was less adversarial in his approach, although there were no major breakthroughs in city-suburb relations during his tenure.

Relationships with the Louisiana state government present a mixed picture. New Orleans is the only large city in the state, and as such, it has long been a target for rural legislators. Historically, the Catholicism of New Orleans has been held suspect by the Protestantism of the majority of the state. In recent years, the state has been hard-pressed fiscally, and this has made it even more difficult for New Orleans officials to get what they perceive to be a more equitable allocation of resources from the state. Personal relationships enter the equation, too. For example, Mayor Barthelemy never got along with former Governor Buddy Roemer. In contrast, Barthelemy and Governor Edwin Edwards were often political allies. Both were strong proponents of casino gambling. Still, working relationships between city and state were never smooth, as evidenced by Governor Edwards's refusal to share a fixed portion of casino tax revenues with the city.

The federal government is a major employer in the New Orleans area. There is no single large defense installation, but several smaller defense facilities make the U.S. Department of Defense the area's top employer. The city has had good access to the federal government. Mayor Landrieu was secretary of housing and urban development during the Carter administration, Mayor Morial chaired the U.S. Conference of Mayors, and Mayor Barthelemy headed the National League of Cities. Despite good working relationships with influential members of Louisiana's congressional delegation, the city was not able to benefit much from federal largess during the years of domestic cutbacks after fiscal year 1978.

There are a vast number of boards and commissions not connected to the city that are responsible for important city functions. These include the Orleans Parish School System, Levee Board, Sewerage and Water Board, and the Board of Commissioners of the Port of New Orleans. *Fragmentation* is the characteristic word we use in discussing economic development and metropolitan governance in New Orleans. It is difficult to establish working relationships among a multiplicity of actors with varied agendas.

LOCAL ECONOMY

New Orleans has never had a strong, diversified economy. Historically, the port has been the base of the city's economy. Since the 1960s, the local economy has had a tripartite base: the port, oil and gas and related industries, and tourism. Some observers of the New Orleans economy might add government and government-related employment and health care to any discussion of the economic base. In fact, of the top 10 employers, 3 are government entities. In 1992-1993, the top 10 employers in New Orleans were the U.S. Department of Defense; the Orleans Parish School System; the city of New Orleans; Avondale Industries, a shipyard that is a Fortune 500 corporation headquartered in New Orleans; Tulane University; Entergy Corporation, the local utility; Ochsner Medical Foundation; Schwegmann Supermarkets; South Central Bell; and Martin Marietta Manned Space Systems, which builds space shuttles for NASA (Mumphrey & Akundi, 1994).

During the 1970s, the New Orleans economy became overly dependent on the oil and gas industries. Whereas there was a total employment increase of 8% between 1969 and 1979, there was a more than 50%

TABLE 4.2 Total Full-Time and Part-Time Employees by Major Industry
for Orleans Parish

Employment	1969	1979	1989
By place of work			
total employment	338,476	366,162	324,561
By type			
wage and salary	313,872	338,742	298,896
proprietors	24,604	27,420	25,665
farm	0	0	0
nonfarm[a]	24,604	27,420	25,665
By industry			
farm	0	0	0
nonfarm	338,476	366,162	324,561
private	284,469	308,482	264,859
agricultural services, forestry, fishing, other[b]	781	701	745
mining	9,151	13,848	13,176
construction	17,784	15,987	7,880
manufacturing	32,696	26,293	16,799
transportation and public utilities	38,785	38,318	26,502
wholesale trade	25,923	26,950	15,895
retail trade	48,568	55,342	46,969
finance, insurance, and real estate	25,664	33,303	25,906
services	85,117	97,740	110,987
government and government enterprises	54,007	57,680	59,702
federal, civilian	12,462	13,631	13,431
military	4,982	5,494	7,425
state and local	36,563	38,555	38,846

SOURCE: Data from Center for Business and Economic Research (1993).
a. Excludes limited partners.
b. "Other" consists of the number of jobs held by U.S. residents employed by international organizations and foreign embassies and consulates in the United States.

increase in mining employment (see Table 4.2). When oil prices col-
lapsed in the 1980s, the local economy also declined, as evidenced by
an 11% decrease in total employment. During the mid-1980s, the city
had a double-digit unemployment rate for more than three years. The
economy recovered, to some extent, starting in the late 1980s. Almost
all of that recovery has been in services, much of it stemming from the

growth in tourism, which is very much encouraged by city officials. For instance, the number of conventions held in New Orleans rose from 1,092 in 1985, with 770,000 delegates, to 1,360 in 1989, with more than a million delegates. More than 11 million tourists visited New Orleans in 1989 (Brooks & Young, 1993, p. 262). Thus unemployment rates are now more in line with national averages.

CITY REVENUES AND EXPENDITURES

New Orleans's primary revenue source is sales tax revenues. In 1991, they accounted for almost 40% of city revenues. The sales tax is high, with a current rate of 9%. The sales tax rate has tended to increase since 1965, reflecting the city's need to compensate for the loss of federal and state aid monies since the 1980s (see Table 4.3). Despite a homestead exemption of $75,000 and a system of elected assessors that surely diminishes returns, property taxes and millages accounted for about 17% of city revenues in 1991. Other sources of revenue are a variety of taxes (e.g., beer and wine taxes); licenses and permits, with the New Orleans Public Service Industry (electric utility) franchise being an especially significant source; intergovernmental revenues, such as the state tobacco tax and the state transportation fund; and service charges (or user fees) exemplified by the charge for sanitation services (see Table 4.3).

When city expenditures are examined, it is obvious that city personnel expenditures, especially in the police and fire departments, constitute the largest part of the city's budget. In 1991, police and fire consumed about 40% of the city's budget (New Orleans, 1993, pp. i-xv).

Development/Redevelopment Programs

New Orleans was a rather late entrant into urban renewal and other 1960s-era federal urban programs. The Louisiana State Legislature did not pass state-enabling legislation for urban renewal until 1968. Thus New Orleans was not involved in the nation's major postwar redevelopment effort until 20 years after the program began. Still, federal funds began to have a major impact on the city. During the Landrieu administration, federal funds went from constituting 1.7% ($1.2 million) of the city's operating budget in 1970 to making up 36% ($78 million) of that budget in 1978. The influx of federal dollars reached its peak during the

TABLE 4.3 Categories of Revenue as Percentage of Total Revenues
Generated by the City of New Orleans, 1965-1985

Category	1965	1970	1975	1980	1985
Taxes	54.2	63.2	36.7	34.2	48.6
real estate	11.4	8.9	2.3	3.1	3.6
dedicated millage (fire and police)	7.7	5.1	2.6	1.8	1.9
personal property	6.5	4.8	1.8	1.7	1.7
sales	26.6	39.4	27.7	24.2	34.3
other	1.9	5.1	2.3	3.4	6.2
Licenses and permits	15.3	12.4	8.4	8.2	10.1
Fines, forfeits, and penalties	4.5	4.1	1.5	1.5	2.7
Service charges	3.1[a]	6.5[a]	3.8[a]	12.6[b]	7.2
Intergovernmental revenues	20.1	12.1	42.0	42.1	28.9
Miscellaneous revenues	2.8	1.6	7.1	1.4	2.4
Total revenues	100.0	99.9	99.5	100.0	99.9

SOURCE: Data from Young (1986).
a. Includes the budget categories "revenue-use of money" (minus interest income) and "charges for current services."
b. Includes road use and special real property charges.

early days of the Morial administration, but then declined to 8.2% ($24.3 million) of the operating budget by 1986. Federal dollars stabilized around this amount (i.e., 8%) during the Barthelemy administration (Bureau of Governmental Research, 1992, pp. 15, 29, 46).

DOWNTOWN REDEVELOPMENT EFFORTS

The Landrieu Administration

After Moon Landrieu took office in 1970, New Orleans was quick to take advantage of federal funding opportunities for downtown revitalization. Some of the earliest efforts could most politely be described as learning experiences. An eight-block area of the largely black, historically significant Treme neighborhood was razed for the construction of a cultural center complex. Later, Armstrong Park, patterned after Tivoli Gardens in Europe, was developed within the cultural center complex as a memorial to Louis Armstrong. Federal urban renewal funds were involved in the $10 million project. The architecturally acclaimed Piazza d'Italia was intended as a mixed-use development that would recognize the city's Italian heritage. Federal Economic Development Administration (EDA) funds were used for this project. The site now consists of a desolate park space surrounded by urban wasteland and debris, as no further development has occurred on the site. Neither the Piazza d'Italia nor Armstrong Park ever achieved the goal of economic stimulus set by the city.

A more successful project was the Superdome, which was developed and constructed while Moon Landrieu was mayor. A megastructure built to serve as a convention/sports arena, the Superdome is located on approximately 52 acres of land immediately adjacent to the city's central business district. With this location, the facility became an anchor that spurred office and hotel construction in its vicinity, while serving as a tourist attraction. The Superdome was completed at an approximate cost of $163 million. Using no federal funds, it was financed through the issuance of bonds by the Louisiana Stadium and Exposition District, a state agency. The bonds were secured by a hotel-motel occupancy tax levied in Orleans and Jefferson Parishes. The city has no direct responsibility for the cost of the Superdome or for its operating expenses (Bureau of Governmental Research, 1992, pp. 19-20).

The Superdome opened in 1975. Its proponents say that it was a catalyst in generating high-rise office construction along Poydras Street, and that it sparked the rejuvenation of the central business district. On the negative side, one can say that it seems successful only as a facility for professional football and for certain "megaevents," such as national political conventions. Its failure as a convention facility is evidenced by the construction of an entirely new convention center in the 1980s. Its failure as an all-purpose sports facility is manifest in the current plans to build an indoor sports arena downtown and an outdoor baseball stadium in suburban Jefferson Parish (for more discussion on this issue, see Smith & Keller, 1986; Whelan, 1989, pp. 229-230).

Another lasting Landrieu administration achievement was the creation of the Downtown Development District (DDD). The DDD is a special taxing district that taxes downtown businesses and uses the proceeds to maintain the viability of the central business district. The DDD provides extra services such as police, sanitation, and cultural events, and provides capital improvements such as new sidewalks and pedestrian malls. The DDD is the outgrowth of a recommendation of the Growth Management Program (GMP) adopted by the city in 1975 (Wallace, McHarg, Roberts, & Todd, 1975, pp. 38-40).

The city and the Chamber of Commerce believed that a growth management program was necessary because of the rapid growth occurring in the central business district and because they wanted to forestall the mounting protests of preservationists who perceived deleterious effects of this growth on the French Quarter (which is adjacent to the CBD) (see Brooks & Young, 1993). The city-chamber steering committee retained Wallace, McHarg, Roberts, and Todd, a Philadelphia architecture and planning firm, as consultants. The consultants called for high-intensity development near previous concentrations in the Poydras/Riverfront corridor. General goals of the GMP included strengthening downtown New Orleans as the administrative office, retail, and entertainment center of the region; promoting growth while preserving historic continuity; returning to the riverfront, with preference given to pedestrian use of the riverfront; providing a full range of activities downtown to attract residents as well as visitors; and forging a public-private partnership to carry out the GMP (Brooks & Young, 1993). Perhaps the most important goal that was implemented was the development of a public-private partnership, as evidenced by the creation and work of the DDD.

Riverfront development projects proceeded as part of the downtown revitalization strategy. The first project of this sort, the Hilton Hotel, was completed in 1977, without the use of public funds. The developer, Lester Kabacoff, leased the Poydras Street Wharf from the public port authority for development purposes. This was the first nonmaritime use of CBD riverfront land and is considered a key in paving the way for later riverfront projects. Later developments on the Poydras Street Wharf have included a cruise ship terminal, a high-rise residential condominium building, and a riverboat casino.

A controversial mixed-use commercial project begun during the Landrieu administration was Canal Place, located at the foot of New Orleans's main downtown thoroughfare adjacent to the Mississippi River and straddling the CBD and the French Quarter. How this project would affect the French Quarter was a major point of contention between the developer and the preservationists. In an effort to resolve potential conflict, the developer started holding open planning forums as early as 1973, and subsequently changed the design of the project substantially. Still, the preservationists were not satisfied, so when the city applied for an Urban Development Action Grant (UDAG) on behalf of the developer, the preservationists used the requirement for an impact assessment to press their case (Brooks & Weeter, 1982). In the end, HUD released the UDAG money for the project, but attached a number of conditions in an attempt to safeguard the French Quarter. The $6 million UDAG was used for infrastructure improvements in the area. Half of the money was to be repaid to the city by the developer. The project was completed under the Morial administration.

THE MORIAL ADMINISTRATION

In any discussion of downtown revitalization during the Morial administration, the 1984 World's Fair stands out as a central event in downtown and riverfront development. The fair itself ran from May to November of 1984 on a site that consisted of 82 acres on the riverfront adjacent to the central business district. The fair had a financial deficit of $100 million, partially because of overly optimistic attendance projections. However, in classic New Orleans fashion, everyone enjoyed the fair ("Laisser les bon temps rouler"). More important, it has been credited with major residual benefits, especially in transforming an area of wharves and warehouses into an area of tourist attractions (Tews, 1986). According to Tim Ryan, University of New Orleans economist, "What the fair did, clearly, was to open up the riverfront area to the tourism industry, and that had never been the case [here] before. A lot of activity occurred after the fair was over" because of the improvements made to the area in preparation for the fair (quoted in McClain, 1994; reprinted by permission).

Chief among those residual benefits was the New Orleans Convention Center. Constructed as the main building for the World's Fair, it became the Convention Center in 1985. The original 15-acre building (820,000 square feet, of which 381,000 square feet is exhibit space) was built with $17.8 million in federal funds, $30 million in state funds, and $48.5 million in Exhibition Hall Authority revenue bonds supported by a 2% hotel-motel tax. After an expansion completed in 1990 (including an additional 330,000 square feet of exhibit space and 7,000 square feet of meeting rooms), the Convention Center is now one of the three largest convention halls in the United States. The Convention Center and the convention industry clearly constitute an arena in which New Orleans is a major-league city. In 1993 the Convention Center held 104 events with a total of 405,334 participants. More than 23,000 jobs in the local economy can be attributed to the activities of the Convention Center.

TABLE 4.4 Morial Convention Center Nine-Year Economic Impact and Tax
Revenue Generated (in millions of dollars)

	1985	1986	1987	1988	1989	1990	1991	1992	1993
Spending									
direct	223.7	241.8	254.5	283.6	466.0	505.9	699.4	801.7	756.8
secondary	335.6	362.7	381.8	425.4	447.1	481.9	725.0	806.2	1,153.9
total	559.3	604.5	636.3	709.0	913.1	987.9	1,424.4	1,607.9	1,910.7
Taxes									
state	24.4	27.4	28.6	34.2	44.1	49.3	58.1	67.9	54.9
local	16.9	18.1	19.2	22.2	29.8	33.1	41.2	50.6	42.6
total	41.3	45.5	47.8	56.4	73.9	82.4	99.3	118.5	97.5
Nine-year total economic impact			$9.30 billion						
Nine-year tax revenue generated			$652.8 million						

SOURCES: Data from McClain (1994) and Convention Authority of New Orleans (1993).

The total nine-year economic impact of the Convention Center has been estimated at $9.3 billion (see Table 4.4). The Convention Center is slated for another expansion in the mid-1990s; it is expected eventually to house more than a million square feet of exhibit space.

During the first Morial administration, New Orleans enjoyed a construction boom in the central business district. Office space more than doubled, from 7.7 million square feet in 1975 to more than 16 million square feet in 1985. One Canal Place, the mixed-use project mentioned above, was begun in 1977 and completed in 1983. Between 1978 and 1981, seven major office buildings were constructed downtown, including the Pan American Building, the 1250 Building, the Exxon Building, Chevron Place, 1515 Poydras Street, and the Poydras Center. Much of this construction was related to the prosperity of the oil and gas industry during this era. The coming of the World's Fair also stimulated some of this construction, especially in the hotel industry. Major new hotels, such as the Sheraton and the Intercontinental, began operating during the early 1980s. The number of hotel rooms increased from 10,686 in 1975 to 19,500 in 1985 (Brooks & Young, 1993, p. 262). Public sector involvement in these downtown projects was largely limited to facilitating development. However, One Canal Place and the Sheraton Hotel benefited from UDAG funds that provided direct loans to the developers.

Another direct offshoot of the World's Fair was the Riverwalk development, which opened in 1986. Riverwalk is a Rouse Company festival marketplace development. The upper level was originally built as the International Pavilion for the World's Fair. The lower level occupies part of the Poydras Street Wharf and Spanish Plaza. This $55 million project also received an $8 million low-interest loan through the UDAG program. Several years of complex negotiations among the Dock Board, city government, and private interests took place before this project started. Approximately 2,300 people are employed in the 180,000-square-foot Riverwalk. The 200 shops and restaurants in the development attract 23,000 visitors on an average day (New Orleans, 1987). Riverwalk general manager Bryan Laide has said that the Riverwalk is one of the top 10 performing shopping centers, in sales per square foot, of the 72 Rouse developments. He expects it to climb into the top 5 during 1994 (quoted in McClain, 1994).

The development of the Warehouse District is another direct impact of the World's Fair. In 1977, 76 acres were designated as a historic district. In a negative sense, it is true that World's Fair development drove some businesses out of the district because of land speculation and fears of traffic congestion. This is hard to document, because some were leaving anyway owing to changes in their space needs in the modern era. Therefore, many believe the times were changing for the Warehouse District even without the World's Fair. On the other hand, it is clear that the fair accelerated the development of the district as a residential neighborhood and cultural area. Within 10 years after the fair, at least 10 historic buildings were renovated for residential purposes. As of 1992, there were some 1,036 apartments with 2,500 residents (Bureau of Governmental Research, 1992, pp. 42-43).

Some of this renovation occurred in buildings used for the World's Fair. Much of the renovation was financed through tax credits offered by the federal government for renovation of historic buildings. The Warehouse District has also evolved into a cultural center, with 13 art galleries and several restaurants and nightclubs. The city's role in the development of the Warehouse District included adding new streets and sidewalks in the area before the World's Fair opened. The location of the Convention Center on the edge of the district also provided another stimulus to the area. In addition, the Downtown Development District promoted the concept of living downtown and provided extra street lighting, a special police substation, and street trees (Bureau of Governmental Research, 1992, pp. 42-43).

Also stimulated by the World's Fair was the development of the multiphase Jackson Brewery complex. The first two buildings of the project constitute a six-level retail and dining complex (135,000 square feet of retail space) on the site of a former brewery on the riverfront in the French Quarter. The first building opened in 1984 and the second in 1986. A third building, separated from the other two by a parking lot, opened in 1988. This building is the most financially successful of the three. Its tenants include the Hard Rock Cafe, Tower Records, and Bookstar. Although the development was privately financed, the city facilitated development by providing the developer with needed parcels of property in a land exchange. The location of this project within the French Quarter caused concern among preservationists. However, because the project did not use federal funds, as in Canal Place, the preservationists could not appeal to the federal government for assistance concerning an impact assessment. The project was approved by the city council in 1982 (Brooks, Wilkinson, & Young, 1984).

Another major downtown project begun during the Morial years was the New Orleans Center, which was completed under the Barthelemy administration. This office and retail complex was developed on land adjacent to the Louisiana Superdome. Given that the city and state owned part of the land, the Morial administration negotiated a land swap between the city and the developer, the DeBartolo Corporation (a national leader in shopping center development and professional sports). The retail component opened in 1988; the office tower opened in 1990.

THE BARTHELEMY ADMINISTRATION

During Sidney Barthelemy's mayoral years (1986-1994), continued development of the riverfront was the major emphasis in the city's downtown development strategy. The major achievement was the construction of the Aquarium of the Americas and Woldenberg Park. In November 1986, New Orleans approved a $25 million bond issue to fund the development of an aquarium and riverfront park. Along with the bond issue money, the Audubon Institute (a private nonprofit organization that manages the zoo and the aquarium) raised another $15 million in private contributions. The aquarium and park opened in 1990 on a 16-acre site fronting the river at the foot of Canal Street directly adjacent to the Riverwalk. The aquarium has been very popular, with attendance exceeding original projections. The park provides riverfront access and an attractive setting for special events.

The riverfront emphasis received further impetus from Riverfront 2000, a long-range "plan" for the development of the downtown riverfront proposed by Mayor Barthelemy and the Audubon Institute in 1990. This is not really a plan in the formal sense of the word, but a proposed series of projects along the riverfront, such as a second phase for the aquarium, expansion of Woldenberg Park, an insectarium, and a natural history museum. Riverfront 2000 also includes the construction of a new facility for the New Orleans Center for Creative Arts (NOCCA), a high school for talented young artists and musicians. These proposed projects would be funded through a combination of city and state bond issues, a hotel tax, and private donations (Bureau of Governmental Research, 1992, p. 60). The Barthelemy administration worked hard to convince the State Legislature to provide funding for an additional expansion to the Convention Center, which would make it one of the largest in the world.

The Riverfront Streetcar Line opened in August 1988. The streetcar runs along the river from Esplanade Avenue (the end of the French Quarter) to the Convention Center, a distance of 1.5 miles. This $5 million project was financed by a public-private partnership involving the Regional Transit Authority, the Downtown Development District, the city of New Orleans, and the Riverfront Transit Coalition, a private nonprofit group made up of property owners along the streetcar line. The streetcar has been very successful, with an average of 4,500 passengers a day (Bureau of Governmental Research, 1992, p. 63).

Mayor Barthelemy and Governor Edwards were both champions of casino gambling as part of a tourism-based development strategy. In 1992, the State Legislature authorized the establishment of a land-based casino on the site of the Rivergate. Located at the foot of Canal Street, the Rivergate opened in the 1960s as the city's first convention center. A permanent casino is expected to open on the site of the Rivergate in the mid-1990s. The choice of this location further emphasizes development (perhaps even overdevelopment) of the riverfront area. In contrast, alternative locations for the casino might have spurred development of other areas in the CBD, other neighborhoods near the CBD (as opposed to the French Quarter), or different parts of the city.

In a related development, the State Legislature authorized riverboat gambling in 1991. By mid-1994, one riverboat gambling boat was in operation at the Poydras Street Wharf. Several other gambling boats on the riverfront are expected in the near future. Again, this decision brought more economic activity to the riverfront area, although in this case there is competition from a gambling boat on Lake Pontchartrain at the other end of the city. Ultimately, the gambling boats will have to compete with the land-based casino.

Downtown development can also be observed in the revival of Canal Street, New Orleans's main shopping street. One large project that has been approved is the conversion of the old D. H. Holmes department store into a complex consisting of a 243-room hotel, 88 apartments, and more than 60,000 square feet of commercial space. The project is to be funded through a combination of federal tax credits (including historic rehabilitation tax credits), a HUD Section 108 loan, and funds from the city's Economic Development Fund (created in 1991 as a result of voter approval of a plan to refinance a portion of the city's long-term debt, resulting in approximately $3.2 million per year devoted to economic development activities). In addition, vacancies are scarce in Canal Street storefronts, and it might be argued that riverfront development is stimulating activity along Canal Street.

During the Barthelemy years, there was a growing awareness of the importance of the health care industry in the local economy and the role it can play in the revitalization of downtown. Such major medical facilities as Tulane University Hospital, the Louisiana State University Medical Center, the Veterans Administration Hospital, and University Hospital (formerly Charity Hospital and Hotel Dieu Hospital) are all located downtown in a corridor along Tulane Avenue. All of these institutions have expansion plans. The Downtown Development District created a medical industry task force in 1990 to develop a strategic plan for the medical institutions. The city's 1992 strategic plan for economic development calls for upgrading or replacing the present physical facilities of the state medical centers, and encouraging the development of the medical center area downtown (Mayor's Strategic Committee, 1992, pp. 110-120).

A final part of the city's downtown strategy is the sponsorship of major sporting events. There are several parts to this strategy. First, the city has one major league sports franchise (pro football's New Orleans Saints) and hosts a major college bowl game (the Sugar Bowl) and a major college football event (the Grambling and Southern University Bayou Classic) every year. The city will make significant efforts and concessions to retain all of the above. Second, New Orleans hosts major sporting events, such as the Super Bowl and the "Final Four" NCAA college basketball tournament. The city will bid on these events in the future, although neither the NFL nor the NCAA may want to hold championship games in a city with legalized casino gambling. These events are all held in the Superdome. Third, the city will try to attract new events. An example is the renovation of Tad Gormley Stadium in City Park for $5.2 million. Gormley was the site of the 1992 U.S. Olympics track and field trials and the 1993 NCAA track and field finals. Finally, there is the proposed sports arena. Approved by the legislature in 1992, it is likely that a new arena will be built adjacent to the Superdome in the mid-1990s. When the State Legislature agreed to assist adjacent Jefferson Parish in building a minor league baseball stadium in 1992, Orleans legislators insisted on an arena for New Orleans. The only question is, Why? At the time the legislation was passed, there was no professional basketball franchise. Without a professional basketball team, there is no prime tenant (and hence no rationale) for a new arena.

CONTAINMENT OF LOW-INCOME POPULATION

A political component to New Orleans's downtown development strategy has been the containment or elimination of "residual low-income populations" adjacent to the central business district and specific development projects (Goetz, 1992). Redevelopment efforts have been plagued by the problems of trying to contain these populations. Western expansion (upriver), after the success of the Warehouse District redevelopment activities, has been constrained by a low-income black population anchored in the St. Thomas public housing complex; northern expansion has been constrained by the Bienville and Iberville public housing complexes and the historically low-income black Treme neighborhood (see Cook & Lauria, 1995); and eastern (downriver) expansion has been historically constrained, first by the French Quarter itself and by preservation interests that have attempted to protect it, and second by the low-income and black Marigny neighborhood. This downriver barrier was removed in the late 1970s and early 1980s when the burgeoning gay population, unable to compete for expensive French Quarter properties, spilled over into the triangle portion of the Marigny and proceeded to gentrify that area (see Knopp, 1990a, 1990b). These gentrification efforts can be viewed as the necessary precursor to the proposed downriver Riverfront 2000 projects and the NOCCA relocation.

The containment or elimination of the Iberville public housing complex has been seen by development interests as the key to the successful redevelopment of Canal Street. At first the DDD tried unsuccessfully to contain the negative externalities associated with Iberville and its residents (see Cook & Lauria, 1995). After 10 years, and with the subsequent support of Mayor Barthelemy, the DDD switched to an elimination strategy. In January 1988, the city of New Orleans released a report titled *A Housing Plan for New Orleans* (Rochon & Associates, 1988). The author of the report was Reynard Rochon, Mayor Barthelemy's campaign manager in the 1986 election. The Rochon report made the claim that there were an adequate number of vacant units in New Orleans to accommodate a "reduction of density" in public housing. This euphemism for razing one-third to one-half of all units was quickly understood (Donze, 1988), and the report caused such an uproar in the black community and the academic community that it was quickly shelved. Shortly afterward, the mayor, at the bequest of the DDD, created the Iberville Housing Task Force and Steering Committee to identify strategies and make recommendations for improving the quality of life for Iberville residents and enhancing the economic viability of the surrounding area (Canal and Rampart streets) (Cook & Lauria, 1995).

Many local observers believe that the formation of the Iberville Housing Task Force signaled the beginning of the development's demise. In addition, in a presentation to the task force, Rochon stated that he believed Iberville to be best suited for designation as "elderly and handicapped" housing. Although the mayor attempted to calm the fears of Iberville residents, many still believed that the task force had been created to find a viable way to "change the character" of the development by housing only elderly and handicapped households, instead of families, as is currently the case. Residents contended that this change in resident profile would result in a "reduction of density" at the Iberville site and, ultimately, precipitate the removal of low-income households altogether from the area. In the end, this strategy also failed; the task force produced a plan that concentrated on improving the quality of life for the Iberville residents rather than improving the economic viability of the surrounding area (Iberville Housing Development Task Force, 1989). Neither Mayor Barthelemy nor the DDD ever endorsed the task force's recommendations. With the advent of casino gambling, the mayor found a new opportunity and shifted his strategy. By insisting that the temporary casino be housed in the Municipal Auditorium in Armstrong Park (Bridges, 1993a), the mayor and development interests appear to be banking on the private displacement of the Treme neighborhood adjacent to the French Quarter (Bridges, 1993b).

NEIGHBORHOOD DEVELOPMENT EFFORTS

New Orleans was one of the first cities in the United States to receive antipoverty funds, and the Landrieu administration was very active in participating in these federal programs. Between 1970 and 1973, the city received approximately $9.2 million a year from the Model Cities program. These funds were targeted for three neighborhoods: Central City, the Lower Ninth Ward, and Desire-Florida. The community-based organization responsible for coordinating the use of these antipoverty funds was Total Community Action (TCA), created in the late 1960s to administer federal funds in selected low-income areas of the city. Throughout the 1970s, TCA received approximately $2 million a year in federal funds (Bureau of Governmental Research, 1992). Although TCA still exists, its influence on city government has waned with the drastic decline in public dollars available to support its efforts.

The Morial administration initiated several neighborhood development efforts. These projects reflected both the federal government's increased interest in neighborhoods during the Carter administration (1977-1981) and Mayor Morial's own emphasis on programs aimed at New Orleans neighborhoods. Neighborhood Commercial Revitalization was a federal program initiated during Carter's term. Its purpose was to revitalize targeted neighborhood commercial strips through loan assistance, exterior design programs, management assistance, and infrastructure improvements. The program began in New Orleans in 1978 with the commercial strip along Freret Street, uptown. Later the program included the North Claiborne/St. Bernard shopping district and the shopping strip along lower Magazine Street. Funding was provided by Community Development Block Grants, loan guarantees from the Small Business Administration, and in-kind donation of professional management and design services from the city (Bureau of Governmental Research, 1992, p. 39). Lower Magazine has been the most successful of the participating commercial strips, largely because private dollars followed shortly after the public monies were invested.

The Regional Loan Corporation (formerly called the Citywide Development Corporation) was founded in 1978 during Mayor Morial's first year in office. This quasi-public nonprofit organization provides assistance to small businesses by mixing private sector financing with loans provided through the Small Business Administration and the CDBG program. (Since 1992, the New Orleans Economic Development Fund has also been a source.) One recent report cites the Regional Loan

Corporation as having been involved in 109 projects totaling more than $35.4 million and representing the creation and retention of 2,690 jobs (Bureau of Governmental Research, 1992).

To the extent that the Barthelemy administration adopted a neighborhood strategy, it centered on the need to upgrade the public housing developments in the city and to lessen the number of vacant units within public housing specifically and the city generally. Of the 13,495 public housing dwelling units, 3,513 (or 26%) were vacant as of August 1993. According to the Housing Authority of New Orleans (HANO), as many as 60% of the units in the Desire housing development were vacant. HANO submitted plans to HUD to demolish 660 units at Desire (New Orleans Office of Housing and Urban Affairs, 1994, pp. 64-66). HUD has approved the plans, and HANO is in the process of embarking on a major modernization project, at a cost of $15 million. Since the city took responsibility for housing rehabilitation in 1986, using federal funds, approximately 1,000 owner-occupied dwelling units have been rehabilitated (New Orleans Office of Housing and Urban Affairs, 1994, p. 64). However, many units remain vacant. The 1990 census found that there were more than 37,000 vacant units in the city. The high vacancy rate became a major issue during the 1994 mayoral election, as many community groups felt that the Barthelemy administration had been too slow in responding to this need.

INDUSTRIAL DEVELOPMENT

The Landrieu administration undertook no major industrial development initiative because officials believed that aggressively promoting tourism was the way to enhance the city's economic outlook. This strategy was adopted, in part, because it could be implemented with a relatively unskilled labor force. Conscious of the need to enhance the skill level of the city's labor force before industrial development could take place, the Landrieu administration used federal funds to provide educational and social services to residents of selected poor neighborhoods.

The Morial administration began development of the 7,000-acre Almonaster-Michoud Industrial District (A-MID) in eastern New Orleans. A-MID is a special tax district and is the largest industrial park within the boundaries of a single city in the United States. It was the centerpiece of Mayor Morial's plan to diversify the city's economic base. In a key part of Morial's economic development program, the city committed itself to providing infrastructure improvements and to developing the area as a planned industrial park (Marak, 1982). The Almonaster-Michoud Industrial District is located along the corridor of the Mississippi River gulf outlet. The area lacks both proper drainage and sanitary sewer systems and an adequate road network. A-MID is dependent on city, state, or federal funds to pay for necessary infrastructure improvements. Currently, only one-third of the site has the requisite infrastructure to support development.

Still, much of the city's industrial base was already located in this area, including heavy and light industry and deep-water port users, processing industries, high-technology plants, and trucking terminals. A-MID is the home of the Michoud plant, a federally owned facility that houses the U.S. Department of Agriculture and Martin Marietta's facility for the production of the external fuel tank for the space shuttles. Other industries in the corridor include the Folger Coffee Company, Georgia Pacific, the TANO Corporation (a producer of navigation and energy control equipment), Siemens-Allis Electronics, Litton Industries, and SFE Technologies (a manufacturer of computer parts). Approximately 9,000 people are currently employed in A-MID.

A special property tax assessment on commercial landholders within the district generates approximately $300,000 each year. These funds are used to market the industrial park and to pay the operating expenses of A-MID. In a 1991 citywide election, the district's self-taxing authority was renewed for 10 years.

The Barthelemy administration continued the Morial administration's policies in regard to A-MID. In particular, Mayor Barthelemy was responsible for attracting a $65 million distribution center built by the MacFrugal's Corporation (formerly Pic-N-Save) in A-MID. This huge facility (1.1 million square feet) opened in October 1991. Fewer than 100 people are employed in the center now, although 500 people might be employed if the facility operated at its fullest capacity. To attract MacFrugal's, the city offered a package of incentives. The city provided $2.5 million in real improvements, and the state constructed a railroad overpass that cost $11.5 million. A-MID was also made an enterprise zone, in order to make it more attractive to businesses.

Louisiana created an enterprise zone program in 1985. To be eligible for a tax credit, an employer must recruit at least 35% of its workforce from among the residents of the zone or a contiguous zone. Participating companies may receive a one-time tax credit of $2,500 for each new employee added to the payroll, and state/city sales and use tax rebates on construction materials and machinery and equipment; they may also get additional $2,500 tax credits for hiring individuals who are receiving Aid to Families with Dependent Children. Since the program started in New Orleans in 1986, 94 companies have applied for tax credits. More than 2,800 jobs have been retained, and more than 1,800 new jobs have been created—at a cost of $9.00 in rebates per job (New Orleans Office of Housing and Urban Affairs, 1994, app. 4).

Effects of Programs

First, it seems clear that the major emphasis of federal urban programs in New Orleans has been in the area of downtown redevelopment. Federal monies have been employed to supplement local public and private efforts in central business district and riverfront revitalization, as in the Convention Center, Riverwalk, Canal Place, and the Sheraton. Relatively few redevelopment efforts have centered on the city's neighborhoods since the antipoverty funds have dried up. Those neighborhood efforts have been narrow in scope, particularly when compared with the downtown initiatives. The alarming number of vacant and abandoned properties in New Orleans, combined with the need for housing, strongly suggests the need for a neighborhood agenda. Physical deterioration of neighborhoods exacerbates such problems as unemployment and high crime rates.

Second, the city's emphasis has generally been on rehabilitation, renovation, and preservation of urban space rather than on displacement. There have been three main instances of displacement in the city's history. Two occurred before the period covered by this chapter: displacement caused by construction of the Civic Center and railroad station in the 1950s and displacement along Claiborne Avenue when Interstate Highway 10 was completed in the 1960s. The other major example was in the early part of the era covered here: displacement in the Treme neighborhood for the Armstrong Park project in the 1970s. In each case, low-income African Americans were displaced to complete a public project. Overall, however, the numbers of people moved and the physical amount of displacement in New Orleans have been relatively small compared with other cities. This is especially true in comparison with federal program efforts in other cities. A combination of a ward system of politics and an active preservationist movement has forestalled some of the worst possibilities in this regard.

Third, there is legitimate concern over the future of the French Quarter in its relation to the riverfront development detailed above. A brief listing of new riverfront facilities in the past decade includes the Jackson Brewery, the Aquarium of the Americas, Riverwalk, the Convention Center, and a riverboat for gambling. With a casino soon to be added to the mix, it is reasonable to say that preservationists have real reason to fear. Their concerns include the likelihood that large-scale facilities will simply overwhelm the architectural integrity and character of the French Quarter and the possibly detrimental effects of bringing more thousands of automobiles and people into an area that is already often crowded. Sadly, the city has done little to plan for this eventuality. We are reminded of Yogi Berra's remark about a popular restaurant: "Nobody goes there anymore; it's too crowded."

Policy Implications

In comparison with the other cities discussed in this volume, New Orleans was a relatively late starter in the federal aid process. Baltimore, for example, was one of the first and most active participants in the old urban renewal program; New Orleans began its urban renewal projects at about the time the program was being phased out by the federal government. Indeed, it is well known that federal fiscal aid to localities declined in real terms beginning with fiscal year 1978, during the Carter administration, and in absolute terms beginning with the Reagan years in the 1980s. Thus New Orleans barely began its efforts with federal programs before the monies were cut back, truncated, or cut off completely. Most of New Orleans's revitalization efforts have emphasized local initiatives. In this regard, it may be more like Canadian cities than like its American compatriots addressed in this volume. There are, however, some policy implications that can be drawn from this effort:

1. Over a period of more than 40 years, federal policies have emphasized the rebuilding and rejuvenation of downtown areas. Federal monies have been used to construct the accoutrements of central business districts (stadiums, convention centers, aquariums, and so on). The New Orleans case exemplifies the general point. The projects mentioned above (e.g., the Superdome, Canal Place, the Convention Center) are all

downtown-oriented projects. In New Orleans, there are no comparable success stories in neighborhood revitalization. We suggest that it is time for federal urban policy to have a "tilt" or redirection in favor of urban neighborhood revitalization. We have had lip service to urban policies with a neighborhood emphasis before: after the 1968 amendments to the Housing and Urban Renewal Act, and during the Carter administration, when HUD established a neighborhood office. New Orleans neighborhood organizations with national linkages, such as LISC (Local Initiatives Support Corporation) and ACT (All Congregations Together) have been visible and active at the local level in the past few years. At the national level, the new "empowerment zone" program can be the vehicle for a redirection toward neighborhoods. Given our experience, the implementation of empowerment zones must be monitored carefully.

2. For many years, the federal government played a direct role in the comprehensive planning process at the local level. In smaller cities, this occurred through Section 701 (of the Housing Act of 1954) planning grants. In larger cities, such as New Orleans, the Model Cities program provided planning monies. These programs were discontinued in the 1970s and 1980s because of federal cutbacks. The research presented in this chapter indicates that there is a diversity and multiplicity of plans operating in the New Orleans area. Perhaps there is, again, a need for federal monies for the development of comprehensive plans at the local level. In today's lexicon, such plans will be called "strategic plans" or "policy plans." Federal money might serve as a catalyst for the coordination of diverse local plans.

3. Our discussion of public housing above has some major policy implications. The research indicates that diverse strategies are necessary, and that such strategies should be chosen in conjunction with the affected populations. In the case of the Iberville housing development located in close proximity to the CBD, the razing of units is unacceptable to the residents. However, this area needs planning that integrates the residents with increased economic opportunities in the area through casino development and the development of the CBD and the French Quarter. In the case of the massive Desire housing development located in an isolated area of the city, razing some units is an acceptable strategy. At the same time, rebuilding public housing can be an activity that provides new skills and new careers for public housing tenants.

References

Bridges, T. (1993a, October 2). Temporary casino site a bitter fight. *New Orleans Times-Picayune*, pp. B1-B2.

Bridges, T. (1993b, October 10). Roulette on Rampart. *New Orleans Times-Picayune*, pp. A1, A6.

Brooks, J. S., & Weeter, D. (1982, July). Canal Place: A clash of values. *Urban Land, 41*, 3-9.

Brooks, J. S., Wilkinson, T., & Young, A. H. (1984, June). The Jackson Brewery: Resolving land use change in a cooperative mode. *Urban Land, 43*, 20-25.

Brooks, J. S., & Young, A. H. (1993). Revitalizing the central business district in the face of decline: The case of New Orleans, 1973-1993. *Town Planning Review, 64*, 251-271.

Bureau of Governmental Research. (1992). *Inventory of economic development programs and strategies in the city of New Orleans, 1970-1992* (Working Paper 2). New Orleans: University of New Orleans, College of Urban and Public Affairs, National Center for the Revitalization of Central Cities.

Center for Business and Economic Research, Regional Economic Information System. (1993). *Full-time and part-time employees by major industry for counties and metropolitan areas (number of jobs)*. New Orleans: University of New Orleans.

Convention Authority of New Orleans. (1993). *Morial Convention Center* [Brochure]. New Orleans: Author.

Cook, C. C., & Lauria, M. (1995). Urban regeneration and public housing in New Orleans. *Urban Affairs Quarterly, 30*(4).

Donze, F. (1988, February 18). Housing project residents blast relocation proposal. *New Orleans Times-Picayune*.

Goetz, E. G. (1992). Land use and homeless policy in Los Angeles. *International Journal of Urban and Regional Research, 16*, 540-554.

Harrison, B., & Bluestone, B. (1988). *The great U-turn: Corporate restructuring and the polarizing of America*. New York: Basic Books.

Hirsch, A., & Logsdon, J. (Eds.). (1992). *Creole New Orleans: Race and Americanization*. Baton Rouge: Louisiana State University Press.

Iberville Housing Development Task Force. (1989). *Report of the task force*. New Orleans: Downtown Development District/City of New Orleans.

Knopp, L. (1990a). Exploiting the rent-gap: The theoretical significance of using illegal appraisal schemes to encourage gentrification in New Orleans. *Urban Geography, 11*, 48-64.

Knopp, L. (1990b). Some theoretical implications of gay involvement in an urban land market. *Political Geography Quarterly, 9*, 337-352.

Marak, R. (1982). The biggest industrial park of all. *Planning, 5*, 10-12.

Mayor's Strategic Committee for Economic Development, New Orleans. (1992). *A blueprint for economic revival*. New Orleans: Author.

McClain, R. (1994, May 12). Big losses paved way for growth. *New Orleans Times-Picayune*, pp. A1, A8.

Mumphrey, A. J., & Akundi, K. (1994). *City of New Orleans: A statistical chartbook 1960-1990*. New Orleans: University of New Orleans, College of Urban and Public Affairs.

New Orleans, City of. (1987). *Overall economic development plan, 1987-1990.* New Orleans: Author.

New Orleans, City of. (1993). *Operating budget for calendar and fiscal year 1993.* New Orleans: Author.

New Orleans Office of Housing and Urban Affairs. (1994). *Comprehensive housing affordability strategy: 1994 annual update.* New Orleans: Author.

Rochon, R., & Associates. (1988). *A housing plan for New Orleans.* New Orleans: City of New Orleans.

Smith, M. P., & Keller, M. (1986). Managed growth and the politics of uneven development in New Orleans. In S. S. Fainstein, N. I. Fainstein, R. C. Hill, D. Judd, & M. P. Smith, *Restructuring the city: The political economy of urban redevelopment* (pp. 126-166). New York: Longman.

Tews, M. K. (1986). *The evolving cityscape: Socio-economic impacts of the 1984 Louisiana World Exposition.* Unpublished master's thesis, University of New Orleans, Urban and Regional Planning Program.

U.S. Department of Commerce. (1993). United States summary: Population and housing unit counts. In U.S. Department of Commerce, *Statistical abstract of the United States 1993.* Washington, DC: Government Printing Office.

Wallace, McHarg, Roberts, & Todd. (1975). *Central Area New Orleans Growth Management Program 1975: Technical report.* Philadelphia: Author.

Whelan, R. K. (1989). New Orleans: Public-private partnerships and uneven development. In G. Squires (Ed.), *Unequal partnerships: The political economy of urban redevelopment in postwar America* (pp. 222-239). New Brunswick, NJ: Rutgers University Press.

Whelan, R. K., Young, A. H., & Lauria, M. (1994). Urban regimes and racial politics in New Orleans. *Journal of Urban Affairs, 16,* 1-21.

Young, A. H. (1986). The revenue picture in New Orleans. In A. H. Young (Ed.), *The state of black New Orleans* (pp. 12-21). New Orleans, LA: The Urban League of Greater New Orleans.

5

Central-City Revitalization

The Fort Worth Experience

ELISE M. BRIGHT
RICHARD L. COLE
SHERMAN M. WYMAN

Over the past 30 years, there have been numerous neighborhood and downtown redevelopment projects proposed and initiated in the Fort Worth, Texas, area. These projects have varied greatly in their stated goals, locations, availability of funding sources, levels of public and private support, scales, and many other factors. In this chapter, we summarize five large-scale revitalization efforts in the Fort Worth area, all of which received considerable federal funds, that represent a cross section of both project characteristics and revitalization results. For each project, we identify those elements that contributed to successful redevelopment and those that inhibited success.

Methodology

The five areas selected for study are known as downtown Fort Worth, the Fort Worth Stockyards, Magnolia Avenue, Polytechnic Main Street,

128

and Regalridge Square. In this report, for each project we identify the project's goals, major participants, type of organization used, and implementation strategies employed. We also evaluate each project's degree of success and try to identify factors that we believe to be associated with its success or failure. Data were collected from three main sources: interviews with notable public and private figures familiar with each project, published material about each project, and statistical data such as those available from the 1960 to 1990 censuses.

The interviewees were public officials and private citizens who were directly involved in each revitalization project. The interviews provided data relating to all five research objectives listed above, and they were the main sources of data relating to some of the first four objectives. Interviewees were asked to summarize the goals, expected benefits and beneficiaries, and impetus behind each project; to describe the roles of all principal players, significant funding sources and other economic development tools utilized, and chain of command during project development and management; and to evaluate the project's success in terms of its objectives, benefits to surrounding areas and the city as a whole, and crucial elements contributing to success or failure.

In order to evaluate project success we relied on both the subjective evaluations of community leaders and objective data, including percentage of families living below the poverty level, unemployment rates, population change rates, and median family income. For each of these, census data from past decades were analyzed for the primary census tract in which the project was located, the tracts immediately adjacent to it, and the city as a whole. We based our final determination of success or failure on these data, the opinions of those interviewed, and the degree of progress made toward realizing the stated goals of the project.

City Context

Fort Worth was established 140 years ago as a small military outpost on the Texas frontier; it soon evolved into a center for livestock trade and processing. In the 1920s, the discovery of three oil fields within 150 miles of Fort Worth fueled population and industrial growth in the area. After World War II, defense-related activities broadened the city's industrial base. The establishment of General Dynamics in 1942, Bell Helicopter in 1951, and Carswell Air Force Base in the late 1950s provided the stimulus for an increase in all types of manufacturing, trade, and service employment. But although the long-term presence of these major employers was a stabilizing force in the county economy, their location outside the central city probably contributed to the need for its revitalization.

The closing of the north-side meatpacking plants in the 1960s caused central-city employment to decline dramatically. This, coupled with white flight to suburban defense-related jobs, beginning in the 1950s and 1960s, contributed to population decline in many of the older residential and neighborhood commercial areas of the city, including four of the five case studies included here (Fort Worth, 1977).

Recently, post-1990 defense cutbacks have further contributed to citywide economic decline; all sectors have been affected, but particularly real estate, construction, retail trade, and financial services. From 1991 to 1992, the average unemployment rate rose from 6.5% to 7.8%; rates were 15.9% for African Americans and 11.3% for Hispanics (Texas Employment Commission, 1992).

URBAN CHARACTERISTICS AND TRENDS

Fort Worth is the sixth most populous city in Texas and, according to the 1990 census, the 28th most populous in the nation. In 1960, Fort Worth's population stood at 356,268. From 1970 to 1980 the population declined, largely owing to the migration of white residents to suburban areas. The population rose to approximately 450,000 people in 1990. Together, Fort Worth and Dallas (located only 30 miles away) form a large metropolitan hub, commonly referred to as the Metroplex, with a current population of almost 4 million (North Central Texas Council of Governments, 1992; U.S. Bureau of the Census, 1960-1980).

The racial composition of the Fort Worth population has changed significantly since 1960. The white population declined (although it is still the majority), and that of all other major ethnic and racial groups increased. Adjusted income also rose between 1960 and 1990, although there are significant differences among these groups: In 1990 median per capita income was $16,405 for whites, $7,583 for African Americans, and $6,956 for those of Hispanic origin. Of the 13.6% of the population living below the poverty level in 1990, 43.4% were white, 26.7% were African Americans, 20.5% were Hispanic, and 9.4% identified themselves as "other" (North Central Texas Council of Governments, 1992; U.S. Bureau of the Census, 1960-1980).

The total housing stock increased by an astonishing 55% between 1980 and 1990, largely because of a boom in suburban multifamily housing construction. However, single-family detached housing is the predominant structural form in Fort Worth, constituting 61% of the total housing stock in 1990. Of the total housing stock, 13.5% is deteriorating and 4.5% is dilapidated (North Central Texas Council of Governments, 1992; U.S. Bureau of the Census, 1960-1980).

CITY REVENUES AND EXPENDITURES

Fort Worth utilizes three main sources of revenue: property taxes, which have risen in budgetary importance over the past 30 years; the sales tax, which has declined in relative importance; and revenue from other government agencies, about half of which is in the form of grants from HUD and the EPA.

Since 1980, only "debt service" has shown a significant increase in its budget proportion, rising from 9.7% to 22.7% of total expenses. The growth of debt service can be attributed to inflation, which pushed up the prices of goods and services; the oil boom of the mid-1980s, which fueled construction, population growth, and annexations; and the city's need to maintain and expand infrastructure systems. The subsequent decline in development activity caused by the 1986 drop in oil prices and the national recession prevented Fort Worth's tax base from growing commensurately with the growth in indebtedness (Fort Worth, 1960, 1970, 1980, 1990a, 1991a).

INTERGOVERNMENTAL RELATIONSHIPS

Fort Worth is the seat of Tarrant County government, and many county offices provide employment in the downtown area. Municipal agencies have developed a close working relationship with county agencies. The city is heavily dependent upon federal spending to support the local economy, and many federal and state agencies provide downtown employment (Fort Worth Chamber of Commerce, 1992).

Fort Worth is surrounded by more than a dozen suburban communities, and city actions have often inadvertently supported their growth at the expense of the central city. For example, the city supported construction of an excellent transportation network, and major state and interstate highways now cross the city in all directions. The economic benefits provided by this network accrued to the urban fringe; improved transportation to the suburbs cut off inner-city neighborhoods and contributed significantly to central-city decline. Another example is the new Alliance Airport, a huge industrial/freight facility and associated "new town" being developed by Ross Perot, with federal and city funding, far to the north of the central city. Although the city is depending on the complex to stimulate growth in the 1990s, the question of how many of these residents and businesses will be truly "new" to the city (rather than relocated from central parts of Fort Worth) remains to be answered.

Central Fort Worth has benefited from some intergovernmental efforts. The city has worked closely with several state and federal agencies to achieve the relocation of a major interstate highway that currently bisects the downtown area. Also, the city is cooperating with the transit authority, Amtrak, and several intercity bus and rail companies to develop an intermodal transit center in the downtown area. In addition, some portions of the federal Clean Air Act of 1990 and the Intermodal Surface Transportation Efficiency Act appear to have the potential to be of significant support to central-city revitalization (Jim Wright, former U.S. congressman representing Fort Worth, interview, summer 1993).

Development/Redevelopment Programs

GOALS AND OBJECTIVES

Among Fort Worth's goals since the 1960s has been the preservation and enhancement of the central business district (CBD). In 1960, this 300-acre area was the principal employment center and represented a substantial investment of infrastructure. The trend toward increased suburbanization raised concerns among business and community leaders that the central city, in need of infrastructure improvements and faced with declining population, would not continue to be a viable business and cultural center unless action was taken. Thus, in the 1950s, the firm of Victor Gruen and Associates was hired to create a plan for improving downtown's competitive advantage and Fort Worth's status as a "world-class" city (James Toal, former Fort Worth planning director, interview, summer 1993).

THE GRUEN PLAN

The principal goals of the Gruen Plan, presented in 1956, were to improve public access to the central city and to create a pedestrian-friendly environment that would enhance retail and commercial activities and stabilize neighborhoods in the deteriorating peripheral areas of downtown. To achieve these goals, the plan called for the construction of a loop roadway with six large parking garages along it, landscaped pedestrian areas, a transit system to connect the garages to downtown destinations and eventually to the suburbs, and redevelopment of blighted areas around the CBD through the creation of "organic neighborhoods" with ample parks and commercial activities over the next 20 years (Lawrence Halprin & Associates, 1971).

The Gruen Plan was widely hailed as a bold and visionary scenario for Fort Worth's central-city development. Unfortunately, it was never implemented. With an estimated 1960 cost of $41 million for the loop roadway alone, funding was a major concern and federal urban renewal assistance was essential. In 1967, wary of federal intervention in municipal affairs, Fort Worth chose not to accept federal urban renewal aid. Instead, the city opted to develop its own urban renewal plan, the general thrust of which was to eliminate blighted areas through slum clearance. The decision to reject federal aid was a death blow to the Gruen Plan, and by the early 1960s it had been shelved (Emil Moncivias, Fort Worth Planning Department, interview, summer 1993).

However, many of the plan's goals remained on the public agenda. For instance, one of the goals was to make Fort Worth a destination for tourists and conventioneers. The construction of the Fort Worth/Tarrant County Convention Center in 1968 was a major step in achieving this goal, as was the construction of the Water Gardens (in 1974) adjacent to the convention center. Much of the revitalization of downtown Fort Worth has its roots in this plan (Wayne Snyder, Fort Worth assistant planning director and industrial development coordinator, interview, summer 1993).

SECTOR PLANNING

After rejecting federal aid, the city pursued a policy of slum clearance that was intended to encourage new private housing by opening up large tracts of land for development. At the same time, however, the city was pursuing an aggressive annexation and development approval policy in its suburban areas—a policy that sapped strength from older neighborhoods. A specific program for addressing neighborhood issues was not formulated until the 1970s, when the city was divided into planning sectors and strategies were formulated for each sector (George Human, Fort Worth assistant planning director, interview, summer 1993).

The goal of the sector planning process was to divide the city into smaller parcels that would be more manageable for planning purposes. In accordance with the goals formulated for each sector, the Planning Department targeted specific areas for redevelopment efforts by creating target area plans. Redevelopment plans for several inner-city areas, including Magnolia Avenue, Polytechnic Heights, and the Fort Worth Stockyards, grew out of the sector planning process (Roger Line, Fort Worth city manager, interview, summer 1993).

Downtown Fort Worth

REVITALIZATION EFFORTS

In 1978, the Economic Development Administration (EDA) notified the city of Fort Worth that it had been selected to participate in one of 37 demonstration programs known as Comprehensive Employment Development Strategies (CEDS).[1] One problem identified by CEDS was the lack of quality downtown hotel rooms and the related inability to attract tourists and conventions. Therefore, a major recommendation of CEDS was the construction of a large luxury hotel near the convention center to be developed through a public-private effort. Because the need for additional hotel rooms had been identified in CEDS as a key ingredient in stimulating downtown investment opportunities in Fort Worth, CEDS provided the justification for two related projects initiated with Urban Development Action Grant (UDAG) funds.

GOALS AND OBJECTIVES

The city's participation in the downtown redevelopment projects sprang from a desire to better the downtown office and retail environment and to create jobs—specifically, to reverse the trend of infrastructure decay and loss of commercial enterprises, and to provide more amenities for tourists.

The private sector was also concerned with revitalizing downtown. For example, renovation of the Hyatt Regency Hotel began in 1979.

Woodbine Development Corporation made the decision to invest in this project after its success with the development of the Hyatt in Dallas. Bass Brothers Enterprises' goals in the development of Sundance Square were similar to those of Woodbine. The project was an example of the Bass family's devotion to the goal of revitalizing downtown Fort Worth, but it also represents a long-term financial investment and was set up as a market-driven project.

ORGANIZATION

The city provided minimal funding in the form of salaries for persons who worked on the projects and tax abatements for the redevelopment efforts. Local government leaders coordinated much of the effort, provided rights-of-way, and completed the EDA and UDAG grant applications. In 1978, the Woodbine Development Corporation of Dallas and Bass Brothers Enterprises of Fort Worth requested that the city of Fort Worth submit grant applications for UDAG funding to help finance two redevelopment projects downtown.

MAIN STREET AND GENERAL WORTH PLAZA

The Gruen Plan envisioned a green, pedestrian-oriented downtown. This vision began to be realized in the late 1970s and early 1980s with the Main Street/Sundance Square project and the creation of General Worth Plaza adjacent to the convention center.

The Main Street Improvements Project involved $40 million in private money to reconstruct a 313-room hotel and $12 million in private funds to renovate 12 buildings (some of historic significance) for use as restaurants and retail and office space in the Sundance Square area. This $52 million investment by Bass Brothers was leveraged with a $3 million UDAG for improvements on a nine-block length of Main Street. The Main Street reconstruction included the repair of the brick street surface, sidewalk reconstruction, landscaping, lighting, subsurface infrastructure, and a traffic signal system. Planter boxes and street fixtures were coordinated with existing buildings along Main Street.

The second project, the Hyatt Regency parking garage and plaza, was completed in conjunction with the renovation and expansion of the hotel itself. UDAG funds provided $6 million for construction of the plaza and underground parking garage. The plaza serves as a link between the hotel and the Tarrant County Convention Center and was designed to

harmonize with the Main Street improvements. The $32 million reno-
vation of the hotel and the construction of an adjacent 40-story office
tower was undertaken by Woodbine Development Corporation.

In addition to the UDAG-funded projects, the city also received an
$86,000 grant from the EDA to complete improvements adjacent to the
parking garage plaza. But as these funds proved insufficient, the city
allocated $86,000 in capital improvement bond funds to match the EDA
grant. Improvements included landscaping, street and sidewalk reconstruction, construction of a commemorative granite star at the intersec-
tion of 9th Street and Main, and the addition of flagpoles at the convention
center entrance (Larry Jacobs, economic development representative,
EDA office in Austin, Texas, interview, summer 1993).

DOWNTOWN IMPROVEMENT DISTRICT

In 1981, the Fort Worth Planning Department began updating and
revising its sector plans.[2] An outgrowth of this effort was the creation
of a downtown advisory committee made up of downtown property
owners. This advisory committee was the forerunner of the nonprofit
corporation that became Downtown Fort Worth, Inc. (DFWI).

In 1984, the board of DFWI decided to form a long-range planning
committee to determine the future role of the organization in promoting
downtown development. The strategy that evolved out of these meetings contained three major objectives: development of a management
plan and establishment of an assessment district to implement that plan;
development of DFWI into a proactive agency to guide downtown
revitalization; and development of a communication program for marketing, promotion, and information sharing among downtown businesses. In an effort to refine these objectives and to determine the
feasibility of an assessment district, interviews were conducted with
downtown businesspeople, decision makers, and public agency staff
members. These interviews helped a steering committee to prioritize the
needs of downtown in four areas: landscaping, security enhancement,
promotions, and transportation and parking improvements.

A special improvement district budget and plan of services were
proposed by DFWI in 1985, partly in response to a need for funds to
support the Houston-Throckmorton Transit Spine project (discussed
below). This district was the state's first successful public improvement
district (PID). Under the district plan, businesses petitioned the city council

to assess downtown property owners more in taxes to pay for extra services such as maintenance, landscaping, and additional security. The plan was approved by the city 18 months after it was proposed. In its final form, the district had the support of 60% of the property owners, 83% of the taxable land area owners, and 92% of the land value owners.

In August 1986, DFWI was awarded a three-year contract by the city council to provide services, and the improvement district became operational. The award was granted in part because of DFWI's track record and credibility with the downtown business community. The improvement district generates approximately $750,000 per year. About two-thirds of that comes from assessments to private property owners. The city contributes $80,000 to $100,000 per year. The balance comes from direct payments for groundskeeping, additional security, landscaping, promotional and marketing devices (flags, banners, and so on), and other similar services that were previously provided by the city. Concession sales within the district, including the highly successful Main Street Arts Festival, supplement the budget.

HOUSTON-THROCKMORTON TRANSIT SPINE PROJECT

The Fort Worth municipal bus system had always operated as a hub, with all routes entering downtown and all transfers made in the downtown area.[3] In the late 1970s, in order to alleviate the downtown congestion caused by this system, city planners proposed construction of a bus corridor through downtown. An application was made to the Urban Mass Transit Administration for a Houston-Throckmorton bus spine that would take all of the buses coming into downtown northbound on Throckmorton Street and southbound on Houston Street only. The proposal entailed rerouting traffic on Throckmorton, which was a two-way street, and reversing the direction on Houston, which was a one-way northbound street.

The city appropriated general obligation bond funds to pay for the difference between the cost of the project and the available federal funds. The city then approached DFWI, stating that the project was still in need of funding. DFWI obtained letters of credit from central area banks backing its $189,000 commitment to the project and began exploring ways to repay the funds, leading to the creation of the downtown improvement district discussed above. The project was successfully completed in 1992.

DOWNTOWN RETAIL LOW-INTEREST LOANS

In an effort to stimulate retail activity in the central area, DFWI approached eight downtown banks and the Downtown Retail Council, a merchants association, in the spring of 1991 to propose a low-interest loan program for retail business in a limited area of downtown. The program depended on investors from the private sector to buy $10,000 certificates of deposit with a four-year term and an annual interest rate of 4%. Loans would be made at 6% to eligible businesses, including restaurants, retailers, entertainment facilities, and property owners leasing to qualified tenants. The money could be used to expand an existing business, to relocate into the defined area, as working capital, for leasehold improvements, for exterior property improvements, or for furniture, fixtures, and equipment.

Securing the enthusiastic participation of the banks proved to be easy, but enrolling the initial investors was difficult. Ed Bass's agreement to participate provided the program with the impetus to begin in earnest; Tandy Corporation, a large electronics firm headquartered in downtown Fort Worth, also participated. But even with these two investors, the original goal of a $1 million loan program could not be realized: The program began with approximately $750,000.

Although there were 10 applicants for the low-interest loans during the first year of the program, no loans were granted because the banks felt that the proposed projects were too risky (a condition endemic to many new businesses). Although the program failed, it did demonstrate the willingness of DFWI, downtown lending institutions, and businesses to cooperate in revitalization efforts.

IMPLEMENTATION STRATEGIES

Without the UDAGs, neither the Main Street Improvements Project nor the General Worth Square project would have been completed. Furthermore, if these two projects had not been completed, Woodbine Development Corporation and Bass Brothers Enterprises would not have made significant investments in renovating historic structures in Sundance Square and the Hyatt Hotel.

Significant private sector involvement was also essential. In the case of Sundance Square, it is clear that Bass Brothers' dedication and financial strength made this a successful project (Tyler, 1983). The creation of DFWI allowed landscaping to be maintained, extra security to be provided, festivals to be held, and the PID to be formed; thus this private sector venture was another essential element in the revitalization effort's successful implementation.

Portions of the Sundance Square project have also received local tax abatements. Sundance West, a large building that houses upscale residential apartments, a theater complex, and several retail shops, has received a 100% tax abatement on new value for three years during construction and for 10 years after completion. Under state law, an area that has been designated as a "reinvestment zone" gives local government leaders the authority to develop an agreement with the property owners to exempt all or a part of the value of their property from municipal taxes for a specified period not to exceed 15 years. Private developers also took advantage of local and federal tax credits for the rehabilitation of historic structures in the CBD. The Sundance Square project and the hotel renovation qualified for these credits (William Boecker, manager, Sundance Square, interview, summer 1993).

The Fort Worth Stockyards

REVITALIZATION EFFORTS

The Stockyards area, located on the north side of Fort Worth, has over the past several decades been the focus of much attention, renovation, and investment.[4] Understanding the Stockyards requires an understanding of Fort Worth's rich history. The city's location as a railhead fostered the development of the meatpacking industry, which was central to Fort Worth's early existence. Swift and Armour were huge producers of beef at the turn of the century.

In the 1960s, the meatpacking industry decentralized, allowing smaller cities to take advantage of technological advances that opened the doors to a new trade. Fort Worth consequently lost the industries that provided employment for 10,000 unskilled and semiskilled laborers. The unemployment figure for the Stockyards area skyrocketed to almost 20%. This enabled Fort Worth to designate the area as a "pocket of poverty." As the meatpacking industry left the area, problems of vagrancy and crime arose, along with deterioration of the buildings and infrastructure. The high rate of crime in the area posed a significant threat to local residents and discouraged any possibility of investment or relocation into the area.

GOALS AND OBJECTIVES

The North Fort Worth Sector Plan identified the need for a concentrated effort to bring the Stockyards area back to economic health. The idea to reclaim the Stockyards as a tourist attraction grew out of this plan. The overall goal of the Stockyards revitalization effort was to reverse the drain of resources this area had imposed on the city. Objectives were to eliminate the criminal activity from the community, renovate the historical buildings, upgrade infrastructure, create employment opportunities, promote tourism, and encourage investment in the area.

The first step was an application to the EDA for a $75,000 Stockyards Area Redevelopment Study grant. As part of the study's research, a consortium of EDA staff and local government representatives traveled to Chicago to view the stockyards restoration project that had been undertaken there. Taking its cue from Chicago, the city embarked upon a restoration plan that included a traditional western theme for the Fort Worth Stockyards.

ORGANIZATION

Many people have been involved in the revitalization of the Fort Worth Stockyards since the early 1970s. After 1974, city hall led the way in planning and supporting the entire effort, but in the decade that followed the meatpacking plant closures, the city's involvement was simply reactive. The Stockyards Area Restoration Committee (SARC), a group of volunteer businesspeople, encouraged private investment, provided the local business perspective, and oversaw the authenticity of the restoration projects funded through the city. Elected local and congressional leadership took particular interest in the project and led much of the lobbying effort for federal grant money. Local property owners also helped, as they were concerned with the value of their investments. Each of these groups provided input and leadership as needed, but no clear champion facilitated overall coordination of the project. As a result, the different groups often did not act in concert.

IMPLEMENTATION STRATEGIES

After quickly organizing a restoration effort based on the plan, the city of Fort Worth received a $1.879 million EDA grant in 1975, one of the first EDA grants allocated in the nation. At this point in time, the revitalization of the Stockyards became a major priority for the city. The city council appointed SARC to advise them with regard to the business interests in the Stockyards.

One of the first issues addressed by SARC was crime. This was handled by the placement of a police station within the Stockyards. A public-private partnership was established whereby police officers volunteered their own horses for patrol duty and private money was used to construct stables. This visible police activity had an immediate impact on crime in the area. As the risk of crime lessened, residents and business interests alike experienced a renewed sense of security.

Another facet of the restoration process was the preservation of the historical integrity of the Stockyards. The Fort Worth Coliseum was the first building to be designated for historical renovation. A coalition of the Historical Society, the city of Fort Worth, and SARC provided the impetus for this initial renovation project. The funding was provided by the city because it owned the coliseum. The success of this project led the way for other renovation efforts by private entities.

Fort Worth also directed efforts toward infrastructure improvements. This included repaving streets using antique-style brick and replacing street signage, lighting, and benches using a western motif. Marine Creek, which crosses Main Street, was also improved; the creek bed was dredged, weeds were cleared, the creek walls were lined with rocks, and refuse that had accumulated for years was removed.

Other than private investments, which the city encouraged, the major source of funding throughout the 1970s was from EDA grants. Over the course of 10 years, the total of federal funds invested is estimated to be from $8 to $12 million.

Magnolia Avenue

REVITALIZATION EFFORTS

Magnolia Avenue is the main arterial street connecting Fort Worth's hospital district to the north and the Fairmount National Register Historic District to the south.[5] Once a prosperous residential neighborhood, the mid-south area experienced significant decline beginning in the 1960s. In 1990, Magnolia Avenue was a predominantly African American community. Total population declined between 1980 and 1990 by 23.6%. Some of this decline may be a result of the conversion of residential units to commercial uses, given that the housing stock declined by 15% during the same period. Unemployment in the area is well above the citywide rate (22.7% compared with 7.5%). Median family income increased by 26.3% from 1980 to 1990, but 41% of families lived below the poverty level in 1990, compared with less than 14% for the city.

However, within the past decade, Magnolia Avenue's strategic location and good housing stock have spurred new interest in the area. Its close proximity to the hospital district provides retailing and service enterprises along Magnolia Avenue with a broader customer base than that provided by the immediate neighborhood. Recognizing the need and opportunity for retail/commercial growth in the mid-south, the city of Fort Worth in 1980 designated Magnolia Avenue for a major neighborhood revitalization effort.

The project area had been under study by the Fort Worth Planning and Development Department since the 1970s. Prior to Magnolia Avenue's selection as a focal point for commercial revitalization, the city staff, in cooperation with several business and community organizations, prepared an overall redevelopment plan for the area. When money became available in the form of Community Development Block Grant (CDBG) funds, the project began to move ahead. The primary direction was provided by the Planning Department, with the Public Works Department handling the contracting.

GOALS AND OBJECTIVES

The primary goals of the Magnolia Avenue project were to stem the tide of deterioration in the area and to promote investment and revitalization. A revitalized Magnolia Avenue was also to serve as a bridge between the hospital district and the Fairmount neighborhood.

ORGANIZATION

In 1980-1981, the city began coordination with community leaders to facilitate and support the revitalization. Coordination took place through the Historic Southside Business Association, which was formed in 1981. From this point there was a tremendous commitment from community leaders and public officials to bring about the revitalization of the area. Shortly thereafter, the project was officially designated as the Magnolia Avenue Streetscape Improvement Project by the Fort Worth Planning Department.

IMPLEMENTATION STRATEGIES

Based on the design concept studies conducted by the Planning Department, the first phase of construction began in 1981. The main priority for this phase of the project was the reconstruction of the streetscape, including the addition of sidewalks, curbs and gutters, decorative pedestrian lighting, and landscaping. The only funding allocated for Phase I was $400,000 of CDBG money from the federal government; this phase was completed in 1984. Phase II, completed in 1988, was funded with an additional $400,000 in CDBG money and continued the project for three more blocks. Phase III, which involved spending $600,000 for improvements between Fifth and Eighth Avenues, was included in the 1988-1990 capital improvements bond package, so no federal funds were involved. Phase III construction began in 1989 and was completed in 1990.

Apart from the $1.4 million in CDBG and bond funding, no other incentives, such as tax abatements or the creation of special districts, were utilized in the area. However, since completion of the project the business community has established an economic development district and a historic district organization, Historic South Side, Inc., to promote further development.

Polytechnic Main Street

THE REVITALIZATION EFFORT

The establishment of Polytechnic College by the Methodist Church in 1891 was the impetus for the early growth and development of the Polytechnic Heights community.[6] Later renamed Texas Wesleyan University (TWU), the college continues to be regarded as one of Polytechnic's primary assets. The Polytechnic area developed into a small, self-supporting Texas town, complete with a city hall, fire station, and post office. In 1922 it was consolidated with Fort Worth.

The Polytechnic Heights neighborhood is located in the southeast section of Fort Worth, approximately six miles from downtown; it has a current population of about 6,000 residents. Poly was a prosperous residential area until the end of the 1950s, when the community began to experience decay. Homes fell into disrepair and commercial enterprises left the community. During the boom days of the 1970s, small investors acquired property from the original homeowners. The new investors turned the property into rentals, but when the economy declined, the demand for rental housing vanished; neighboring communities offered better, safer housing at lower rents. Investors found their properties continuously vacant; they stopped maintaining them and making payments. The vacant properties were eventually foreclosed, vandalized, covered with overgrown grass and weeds, and used as drug and gang houses. One in every four homes in Polytechnic Heights was vacant in 1990, and more than 660 homes were turned over to the U.S. Department of Housing and Urban Development.

Between 1970 and 1980, the population declined by 1.6%, but the minority population increased dramatically, from 1% to more than 76%. Some 76% of the families in Poly were headed by females, compared with 59.3% citywide.

GOALS AND OBJECTIVES

The Polytechnic area faced a major threat in 1981 when the administration and trustees of TWU began plans to relocate the college across town into suburban southwest Fort Worth. Enrollment was declining, and parents and students were complaining about the crime rate in the area. The news of the TWU move was a major blow to area residents and business owners. The college had been the cornerstone of the community. The Methodist Church also opposed the move. The church elders, neighborhood residents, and business owners began to organize, conduct studies, and pursue other efforts to revitalize the Polytechnic area and keep the college in the community.

The Polytechnic Strategy Committee, a mayoral task force, was formed to address the problems of the Polytechnic area. In 1984, the committee began developing a series of long-term strategies for the area. The committee concluded that the most pressing problems facing the community were employment, housing, neighborhood networking (particularly with TWU), poor community image, and lack of leadership.

Meanwhile, TWU officials continued to pursue plans to relocate the college. They persisted until the spring of 1985, when it became clear that Texas Wesleyan did not have the funds to purchase new land and finalize the move. As a result, the TWU Board of Trustees adopted a resolution rescinding its previous resolution to move the campus, began a major capital and endowment campaign, and engaged in efforts to determine how the university could most effectively become a catalyst for renewal in the Polytechnic community.

ORGANIZATION

The Poly project received strong support from the city, TWU, and various funding sources, and an endorsement from Congressman and House Speaker Jim Wright. However, community support and leadership vacillated.

With the decision to remain in Polytechnic Heights, the college administration became an active partner with the neighborhood and participated in the Polytechnic Strategy Committee. By July 1985, TWU was heavily involved in the revitalization efforts. College administrators prepared a proposal for matching grant funds and sent it to various local agencies, including the city of Fort Worth, for support and input.

In February 1986, after receiving feedback from the various agencies, the college submitted the proposal to two foundations. Structured Employment/Economic Development Corporation (SEEDCO), an intermediary organization of the Ford Foundation, responded to the proposal. However, before it would agree to make any funding decision on the proposal, SEEDCO wanted a written description of the relationship among all entities involved in the project, details of the commercial revitalization plan, and an itemized budget for the second year of the project.

The city of Fort Worth had informally set aside $175,000 in CDBG funds for the southeast quadrant of Fort Worth, which includes Polytechnic. The strategy committee requested that the city commit $160,000 of the CDBG funds for use in the Polytechnic area. However, the Polytechnic Strategy Committee requested that CDBG money for a revolving loan fund be deferred until the question of whether local banks would be willing to participate in the project (and leverage the CDBG funds) was resolved. When the money was eventually made available in the form of 8% loans, no businesses requested them because of the forms and red tape the loan requests entailed. The project also received funding from a number of other sources, including several foundations and donations from more than 100 local corporations.

The strategy committee also requested that the Fort Worth Planning Department prepare a revitalization plan for the commercial strip, to guide interested parties in its renovation. The city responded to the request by agreeing to hold the $160,000 while the plan was completed. In addition, the city executed a contract with TWU in the amount of $26,500 to operate the Polytechnic Main Street Project for one year.

In late 1986, SEEDCO approved a one-year renewable grant for $26,464 to TWU for support of the project development phase of the Polytechnic commercial revitalization project.

IMPLEMENTATION STRATEGIES

To fund these projects, the project director secured $133,500 in CDBG funds from the city of Fort Worth. The project director and the Fort Worth Planning Department conducted a building inventory, a rehabilitation needs assessment, and a merchants survey to determine what needed to be done. Area merchants were encouraged to participate in property rehabilitation.

In spite of these efforts, by 1988 no new businesses had opened in the area, and few business owners seemed interested in making facade improvements or supporting marketing schemes. At this point, SEEDCO urged the board to become a housing-oriented community development corporation instead of a Main Street project. The reasoning was that the program should revitalize the housing stock, encourage higher-income families to move in, and thus improve the environment for businesses to both stay and locate in Polytechnic Heights. However, the Liberation Community and Neighborhood Housing Services in Polytechnic were already successfully providing housing services to the community. This new goal angered the two housing agencies; they felt that the Polytechnic Main Street Project was "stepping on their toes."

The administration at TWU did not want to get into housing. The board decided to rethink the goals of the program and to vote on the SEEDCO versus TWU requests. Meanwhile, the merchants thought the project was being controlled too much by Texas Wesleyan and decided to move the project office off the TWU campus into a storefront location. Angered by the decision, TWU pulled out of the project and started its own economic development department. After TWU pulled out of the project, the city of Fort Worth would only contribute to the project based on matching funds. When the project could not show proof of the funds, the city pulled out too.

Meanwhile, SEEDCO funding for the Polytechnic Main Street Project also began to decline. In early 1990 SEEDCO funded a $10,000 technical consulting assistance project for Polytechnic Heights merchants. However, SEEDCO ignored previous agreements to reimburse training expenditures and soon was behind on grant payments by more than $17,000. By 1991, SEEDCO was no longer a contributor to the project. Polytechnic Main Street requested and received an emergency grant from the Amon Carter Foundation. However, by 1992 the Polytechnic Main Street Project board had to accept the project director's resignation because of lack of funds.

Regalridge Square

REVITALIZATION EFFORTS

Regalridge Square is a 784-unit apartment complex located in south Fort Worth.[7] It was built in the early 1980s by the Briscoe Development Company. The complex contains a recreation area with a game room, four swimming pools, a Jacuzzi, lighted basketball and tennis courts, jogging and bike trails, playgrounds, laundry facilities, a clubhouse, storage facilities, and beautifully maintained and landscaped grounds. Activities include aerobics classes three times a week, a youth program for ages 7 to 17, and a rental assistance program. In addition, there is a day care center on site at Regalridge. The racial makeup of the residents is 80% African American.

The entire Regalridge complex is enclosed with an electronically controlled security fence, and the safety of residents and their property is enhanced by a 24-hour security patrol staff and television surveillance. Visitors must present identification and be on a visitors list in order to enter the property.

Apartments vary among four floor plans, with rental rates ranging from $295 per month for a one-bedroom, one-bathroom unit to $445 per month for a four-bedroom, two-bathroom unit. To qualify for residency at Regalridge, one must have an income of at least $1,000 a month for a one-bedroom apartment. For each additional bedroom, the resident must earn $100 more a month. Rent must be paid by money order. Regalridge Square also participates in the Section 8 housing program. Rental at Regalridge Square is contingent on being employed for at least six months.

Each tenant is given an orientation consisting of a slide presentation and information on apartment rules and regulations and services available.

Rules prohibit sitting on doorsteps, drinking beer in public, playing loud music, and loitering. Residents are not allowed to work on their cars or wash their cars at the complex. Children are allowed to play only in designated areas.

A bus stops in front of the complex every 20 to 30 minutes. Also, the complex is located close to shopping facilities, churches, parks, and recreational facilities. The elementary school is located a few blocks west, and the high school is just across the street. Tarrant County Junior College is located about a half mile south of the property.

The Briscoe Development Corporation initiated the Regalridge project in 1982. One of the partners was Leonard Briscoe, a minority businessman and former Fort Worth City Council member. The Briscoe Development Corporation, in concert with Briscoe Property Management, had been involved in general real estate brokerage, property management, and real estate consulting.

GOALS AND OBJECTIVES

The original goal was to provide safe, attractive housing for low- and moderate-income families in south-central Fort Worth. The developer also wished to demonstrate that large-scale housing projects designated for low- and moderate-income residents can be maintained as attractive and safe places to live. Some city leaders felt that an important, unstated goal of the project was to make money for the developer.

ORGANIZATION

The Briscoe Company, under the leadership of Leonard Briscoe, approached the city council with a proposal to build low-income housing in Fort Worth in 1982. Support for the project had already been obtained from HUD and members of the Fort Worth city staff. Additional support came from three city council members.

Opposition came from a group of three city council members as well as from residents of some surrounding communities who were worried about possible negative effects from the location of a large-scale project in their neighborhood. The opponents felt that there was an absence of public purpose, that the project was not feasible, and that the financial capability of the developer was questionable.

IMPLEMENTATION STRATEGIES

The first task of the Briscoe Development Corporation was to locate property in south-central Fort Worth. Once the property was acquired, the developer quickly began construction on Phase I of the project. Initially, 392 units were built; another 392 units were completed during Phase II of the project, which was completed in September 1986.

Fort Worth appropriated funding for the "Operation I Will" program at Regalridge Square in 1988. The first goal of the program was to help 99 families receiving public assistance to make the transition from public subsidy to self-sufficiency. The program's objective was to provide better housing and subsidized child care at the same time the parents received educational assistance and job training. A second goal of this program was to assist Regalridge in achieving a positive cash flow and loan repayment.

The project was initially financed by two Urban Development Action Grants totaling $3.5 million. These grants were leveraged by $18.7 million in additional funding from the Trinity Housing Finance Corporation, a public nonprofit corporation, which issued two housing revenue bonds, one for $9.2 million and another for $9.5 million. The private money used to leverage the UDAGs was provided by partners Leonard Briscoe, Judge Clifford David, and Judge Riley Sterns. Housing revenue bonds were secured by the first mortgage lien, and UDAG funds were secured by the second mortgage lien. The revenue bonds were to be paid back at 8.875% on Phase I and 8.75% on Phase II. A grace period of 12 years was provided in the repayment of UDAG loan funds to prevent cash-flow problems so that the commitment to Section 8 existing housing program rental rates could be achieved. During the 12 years, 3% interest would accrue on the UDAG loans.

Under Operation I Will, low-income residents had to agree to go to school and work full-time while in the program. Operation I Will is a demonstration project funded with CDBG money. It is a rental subsidy program with social services support and an educational emphasis for families in Fort Worth. Through Liberation Communities, Inc., the city provided funds to subsidize rental payment for approximately 99 apartments. The overall goal is to assist the client in moving toward self-sufficiency by developing a mutually agreed-on case plan that includes financial assessments, employment maintenance, educational needs, job training or retraining, child care, and transportation.

In 1988, the Community Development Council set aside $622,000 to be used for housing subsidies for the program. The following year, the Fort Worth City Council authorized the city manager to execute a 12-month contract with Liberation Communities, Inc., in the amount of $132,615 to administer the subsidies. Of this amount, $35,000 was for administrative costs, $68,139 was for on-site case management, and $29,476 was to be used for rental subsidies for eligible families. The remaining $489,385 was put into an interest-bearing account to be used for subsidies as the need arose. Also, money was set aside for educational assistance for the clients. The city now allocates approximately $134,000 annually to pay for this program.

The Regalridge Square project has been plagued by many financial complications and legal matters. Originally, Frontier Savings held the mortgage on the property. Frontier subsequently filed for bankruptcy. The FDIC eventually sold the mortgage to Heartland Incorporated, a holding company. During the same period, Briscoe Development filed for bankruptcy, but the reorganization terms call for the company to continue to manage the property. Heartland currently holds the mortgage on Regalridge Square and is attempting to sell the property; however, the city of Fort Worth is opposed to this action because the property provides attractive and secure housing for needy families and because the city has made a commitment to those residents to prevent the complex from becoming a "slum." Also, the city has two liens against the mortgage for the UDAG funds that it loaned Briscoe Development for construction of the project. The mortgage case is currently in litigation.

Program Evaluations

DOWNTOWN FORT WORTH

It is clear that the community has benefited from the downtown improvement efforts. Economic value has been added to downtown Fort Worth, the area has experienced dramatic aesthetic improvement, foot traffic has increased, and there has been a stabilization of the area with an increased tax base. In the case of Sundance Square, the owners in the designated area have reaped the benefits of the aesthetic improvements; now that downtown is more pleasing to consumers, there has been an increase in pedestrian activity. Sundance West, a downtown apartment building built by Bass Brothers Enterprises, opened in 1993 and was so successful that expansion plans are now being made. All interviewees believe that the goals of the project have been met. Jobs have been created and there is a larger tax base for the city. The project has affected the community as a whole and has created a nucleus of new development. Interviewees attribute the project's success partly to the creation of this nucleus, believing that all aspects of the project (retail, multifamily residential, and commercial office space) are interdependent and must be present for the achievement of success.

The project has not been a complete financial success, however. For example, the renovation of the historic Hyatt Hotel incurred costs that were not anticipated in the planning stages, a problem that often arises in the renovation of historic structures. The low-interest loan program was not very successful, perhaps because of the poor rates of return offered to investors and the extreme caution of the banks in approving applicants. Accurate feasibility studies are also essential: When Sundance Square was completed in 1982, the retail stores were occupied by upscale shops that proved unsuccessful; these businesses have been replaced by more moderately priced goods and services. Finally, in a mixed office-retail complex such as Sundance Square, it is crucial that retail space be rented even though office space may be the largest use category, because if businesses see that retail shops are not occupied, they are less likely to rent office space.

Still, the city has been able to halt the process of decline that threatened the CBD in 1960. Today, downtown Fort Worth has 31,500 employees, 920,037 square feet of retail-entertainment space, and 7,858,241 total square feet of office space; the area attracts 3.18 million visitors annually (Downtown Fort Worth, Inc., 1992). Although these statistics are encouraging, there are still problems. The downtown commercial property occupancy rate is 75%, crime is a major concern downtown as well as in other areas of the city, and residential neighborhoods surrounding the core are badly in need of improvements (Fort Worth, 1991b). Funding for further improvements is hampered by budget problems resulting from the economic downturn in the region and the nation.

THE STOCKYARDS

The revitalization that took place in the Stockyards was a resounding success in terms of its initial goals. Beyond returns to the businesses and investors, the project benefited residents on Fort Worth's north side

by providing employment opportunities and stabilizing their community. Population in the area increased by 13.4% from 1980 to 1990 (from
8,165 to 9,258 residents), reversing the 15.9% decline in population
experienced between 1970 and 1980. Similarly, population in the adjacent tracts increased by 2% from 1980 to 1990, compared with a 1970
to 1980 decrease of 10.5%. Median family income increased by 58.9%
from 1980 to 1990, coming close to the citywide increase of 69.3%. In
addition, streets were repaired and storm drainage improved (North
Central Texas Council of Governments, 1992).

The Stockyards became a viable, thriving economic center, attracting
tourist dollars. Experts attribute much of Fort Worth's current tourist
and convention business to the appeal the Stockyards area has for
visitors. Stockyards area businesses and residents have been reassured
by a reduction in crime and by the reversal of the unsavory reputation
the area had developed. Public investment has resulted in renewed
private investment. The Stockyards area is now a place where families
can enjoy activities and feel secure. Property values and local business
confidence have increased along with the tax base, providing additional
income to the city and financial stability in the Stockyards.

MAGNOLIA AVENUE

Everyone interviewed regarding the Magnolia Avenue project agreed
that the trend of decline was stopped and a "development" mentality
was born in the neighborhood. Adjacent areas were organized, stabilized, and enhanced even in a poor economy. The area is now prepared
to become the expansion area for downtown in the next decade. The
project is viewed as a resounding success by both the public and private
sectors, although the socioeconomic benefits as reflected in the census
data may not be obvious for another decade.

As of 1989, the Fort Worth Planning Department estimated that the
public improvements on Magnolia Avenue had generated approximately
$15 million in private investment ($4.5 million in renovation of existing
structures and $10.5 million in new construction). Approximately 30
new businesses have chosen to locate on Magnolia Avenue, including
a professional theater and several restaurants (G. Human, interview,
summer 1993). Many property owners have stated that they located
in the area or improved their property as a result of the public
improvements.

POLYTECHNIC HEIGHTS

Census data indicate few positive changes in socioeconomic factors relative to the Polytechnic Heights neighborhood. The 1990 census revealed that there was a 12.2% decline in population for Polytechnic Heights since 1980. As of that census, one in every three houses was vacant. Although the vacancy rate has risen since 1980, the total number of vacant houses decreased from more than 660 to 549. This decrease is probably a result of the fact that some abandoned houses have been destroyed.

Employment and income characteristics have also changed since 1980. Median family income rose by 24.2% from 1980 to 1989; however, median family income for the city increased by 69.8% during the same period. The percentage of families living below the poverty level has declined significantly, from 47% to 38%; but male participation in the labor force dropped significantly (from 71.6% to 58%), and female participation decreased from 51.8% to 37%. The proportion of white-collar jobs in 1990 was 19%; the proportion of low-skilled jobs was 81% (North Central Texas Council of Governments, 1992).

Between May 1987 and October 1989, 15 businesses located in the Polytechnic Heights commercial strip. In 1987, the first year of the project, five businesses located in the neighborhood, including a theater, a restaurant, Fort Worth Independent School District youth entrepreneurial print shop, a convenience store, and a beauty salon. Three new businesses located in the commercial strip in 1988, and seven new businesses moved there in 1989. However, during the same period, 13 businesses left.

A physical survey of the Polytechnic commercial strip revealed that 16 of the businesses that were present in 1987 are not there today and that the buildings vacant in both 1985 and 1987 are vacant today. In addition, an on-site review of the area revealed that the buildings located at the addresses that were targeted for revitalization on the commercial strip remain in similar or worse condition than was described in 1985.

Thus the Polytechnic Main Street Project can be considered only partially successful. Some new businesses came in and fostered new jobs to replace those that left. The program also produced some new sidewalks that were completed by the city in May 1992, some drug houses were closed, and plans for facade improvements were drawn up. In addition, a free lawn equipment loan program was established in the neighborhood. The project was clearly successful in creating some organization in the community; it also established a relationship between the city and the neighborhood and spurred interest from people of influence. Although the successes seem small, some conditions changed that may not have changed without the project. It is clear, however, that the project failed to achieve its major goals.

REGALRIDGE SQUARE

The 1990 census data paint a mixed picture of the effects of Regalridge. Population in the census tract in which Regalridge Square is located increased by 40.4% from 1980 to 1990. During the same period, the white population decreased from 74.5% to 25.3% and the minority population increased by 198%. Median 1990 family income is $25,347 in the area, only slightly lower than the citywide median family income of $30,967. However, the proportion of families living below poverty level is 22.2% for the primary census tract, whereas the citywide rate is 14%.

In his original presentation to the city council, Leonard Briscoe talked of providing housing for low- and moderate-income families. However, Regalridge did not offer any kind of incentive to moderate- and low-income residents until 1988, when Operation I Will was put into effect. The accomplishments of Operation I Will include enrolling about 25 people in college or GED classes and preparing some residents for home ownership by helping them to open checking and/or savings accounts, purchase cars, obtain job promotions and/or salary increases, improve their credit ratings, reduce their indebtedness, and increase their personal incomes. Families in Operation I Will have been provided the opportunity to learn money management skills; to live in modern, well-secured, affordable housing; to receive regular case management services from a trained on-site human services staff; to be exposed to numerous self-help seminars/workshops; and to receive financial assistance with their education. Supporters say that not only has this program been cost-effective in terms of its operation, but the $4,800 average per family cost over two years is a minimal investment for the achievement of the families.

Those opposed to the project claim that only when faced with financial default and the loss of its several million dollars in grant funds did the city council hastily devise a rental subsidy program (Operation Will) to be carried out at Regalridge. Several hundred thousand dollars were committed to the program, without competition, in a desperate effort to save the project. The rental subsidy program was presented as a "demonstration project" to help the less fortunate, when in reality the primary motivation was to save Regalridge from default and to save the mayor and the city council from the political embarrassment of having to explain their support of such an ill-conceived project. As with most politically motivated projects, millions of dollars of public funds were wasted, and the public in general and the less fortunate specifically were poorly served. The facilitator of Operation I Will indicated that the project has had limited success and has been plagued by ineffective case management, and added that there is no pattern of self-sufficiency shown in Operation I Will, and that only one person has moved out of the program into self-sufficiency.

Conclusions

DOWNTOWN FORT WORTH

Bass Brothers Enterprises, the Woodbine Development Corporation, city leaders, and the federal government all worked together to address a problem found in many central cities. This problem is commonly referred to as the "doughnut effect," whereby inner cities decay at the same time the outer ring becomes more prosperous (Barrett & Green, 1992).

Although all roles were significant, what truly made this project successful was the involvement of the Bass brothers. Besides their participation in the projects discussed here, they have recently brought residents back to downtown with their Sundance West project. Main Street revitalization could not have been a success of any type if the Bass family had not worked to make it so. Ed Bass not only financed and oversaw the direction of Sundance Square and Sundance West, he also personally supervised even the smallest details that contributed to making his vision a reality. His personal and aesthetic daring is as much his contribution to the project as his financial solidarity.

Thus it appears that successful economic development projects require tenacity and clarity of goals. Champions who can support these goals are also very helpful: They can provide financial support, as is the case with the Bass brothers and the Hunt family, or leadership in developing a mandate for redevelopment among the public and private sectors.

THE STOCKYARDS

The success of the Stockyards area restoration stemmed from high participation levels on the parts of various segments of society, including private investors, federal and local governments, and local business leaders. However, no true identifiable champion ever emerged. Rather, clear goals coupled with a "can do" spirit shared by many participants carried the project to a successful conclusion. Still, the fragmented leadership meant that efforts were duplicated and projects were started that were not fully evaluated beforehand.

A key factor in securing funding was the fact that Congressman Jim Wright, Fort Worth's representative, was Speaker of the House. His political connections enabled Fort Worth to secure substantial amounts of funds far ahead of other municipalities. However, owing to allegations of wrongdoing in subsequent years, his name actually became a detriment to the process, making it difficult later to secure federal funds. The original plan called for a minimum of 10 years of federal support to ensure adequate funding; cuts in financing caused by changes of administration were not part of the plan. But Fort Worth had become dependent upon federal funds to carry out capital improvements, and when these funds were eliminated, the city was unable to locate alternative funding sources. Thus some actions called for by the plan remain incomplete today.

The success of the Stockyards project was enhanced by fortuitous timing coupled with a focus on the city's unique heritage. The 1970s saw renewed interest in anything with a western flair. The history surrounding the Stockyards area, including the cattle industry and the cowboys that frequented local watering holes, provided a common theme around which to base the redevelopment efforts. Few other cities can boast of this heritage to the degree that Fort Worth can, but most communities do have some unique aspects that could be promoted.

MAGNOLIA AVENUE

According to all the interviewees, the primary reasons for the success of the Magnolia Avenue project were well-defined goals, private and public support for them, and the concentrated effort that was put forth by both sectors. The project gave hope to the area's residents, and the results were tangible. The only difficulty mentioned was a lack of clear leadership and continuity in city staff responsibility for the project

There were three different city offices involved, and this delayed critical funds that were needed to give the project its initial impetus.

The conditions necessary for success existed within the Magnolia Avenue project. Businesses were in place, but they were in need of a higher-quality environment. Good housing stock was available but in need of renovation. Perhaps most important, the community had a sense of identity. Cooperation between concerned individuals in the public and private sectors made this project successful.

The Magnolia Avenue project continues to have a positive effect outside the immediate project area. The most significant and permanent impact of the project has been to mobilize citizens within the area to help themselves by bringing in new business and by cultivating a continuing working relationship between community leaders and city staff to address the area's needs.

POLYTECHNIC HEIGHTS

Strategic planning is required for successful economic development. A strategic economic development plan process can position a community so that it becomes competitive in attracting prospective businesses. Failure in economic development occurs when the economic planning lacks community participation and ownership and when plans, goals, and objectives are too general. All of these factors contributed to the downfall of the Poly program. First, it suffered from poor management. A great deal of money was used to hold meetings and pay salaries; not much was spent on actually getting things done in the community. After a few years of funding the project, funders were not pleased with its progress and lost interest.

In addition, realistic goals and objectives must be established and pursued throughout the life of a successful project, independent of the specific interests of funding sources. The major downfall of this project came as a result of conflicts between the goals of Polytechnic Main Street and SEEDCO. The original goal of the Poly project was to preserve and revitalize the commercial strip. When businesses were not retained or attracted to the community, SEEDCO wanted the goals to be expanded into the housing area. The board's decision to follow SEEDCO's suggestion split the community and led directly to the loss of Texas Wesleyan's support, which caused the city to abandon the project as well.

Effective neighborhood economic development requires that neighborhood members participate in the creation of the plan, yet many Polytechnic Heights residents were not made aware of revitalization efforts and were not included in the organization of the project. None of the project board members lived in the area. The Polytechnic project was not the brainchild of the merchants and residents, it was initiated by the strategy committee, and some members of the community viewed the project as something that was imposed on them. It was hard to gather enough business participation in the Polytechnic Business Association, partly because most of the businesses were run by one- or two-person teams that included the owners, so even their attending meetings was difficult. But lack of interest on the part of business owners probably had more to do with poor leadership and ambiguous goals than with timing.

REGALRIDGE SQUARE

Few would argue with the worthiness of the goal of the Regalridge Square project to provide low-income housing. However, the project would have been more successful if the city had carefully reviewed plans presented by developers to ensure that they warranted public involvement and met public goals. There also should have been community support and involvement in the project from its inception; effective, well-conceived goals and a long-term commitment by the public and private sectors to fulfill them; and a tracking system to document the project's progress and effectiveness.

Summary and Policy Implications

There was general consensus among those interviewed that the downtown hotel and Sundance Square projects and the Stockyard revitalization efforts have been successful. Although vacancy in the central business district is currently 40% for office space, nighttime entertainment and retailing businesses draw 3 million visitors according to statistics compiled by Downtown Fort Worth, Inc. The Stockyards is also a major tourist destination for Fort Worth. In both cases, the projects were spearheaded by area businesspeople and landowners. The success of both projects is the result of public-private cooperation among interested parties. In each case the goals were reasonably defined and articulated, and there was consensus among parties relative to those goals. Moreover, there has been a long-term sustained effort by the public and private sectors to ensure success in these areas. Federal and municipal funds provided the seed money upon which private investors could build. The federal monies were important symbols of public commitment to each of these projects.

The neighborhood revitalization projects (Magnolia Avenue, Polytechnic Heights, and Regalridge Square) are more difficult to evaluate. Magnolia Avenue appears to have been successful in reversing a trend of decline in the area. However, this may have happened without public assistance through the process of gentrification. Magnolia Avenue was not in a severe state of decline at the time that revitalization efforts were initiated. As in the downtown area and the Stockyards, the streetscape improvements served as a symbol of public commitment and interest in the area. Improvement of the visual aspects of the area encouraged private investors to take risks they may not have taken otherwise. Polytechnic Heights was in a more advanced state of decline at the time revitalization began. Furthermore, there was a general lack of consensus regarding the goals of the revitalization effort. Lack of leadership, community support, and a well-articulated agenda for improvements have stymied efforts to make significant accomplishments in the area. Moreover, there has not been a long-term sustained effort on the part of public and private interests. The Polytechnic Main Street Project was funded for only 3 years, not enough time to realize change in an area experiencing so many problems. The Regalridge Square project could not be considered successful in realizing its goal of providing private housing for low-income residents. Management appears to have been as much a problem as funding. Substantial public subsidies may have been squandered by individuals seeking a profit.

Significant policy implications emerge from study of the Fort Worth projects. Each is directly supported by actual successes and failures experienced in the five projects included in this study:

1. A strategic planning process that includes data collection and analysis, collaborative goal and objective setting, and implementation and monitoring committees should be required for all projects. Such a process would encourage early vesting and collaboration between all parties: project residents, project area leaders, property owners, and local, regional, and federal officials. The involvement of regional federal representatives directly in the strategic planning process makes their expertise available directly to the project. Moreover, such involvement encourages a more

knowledgeable and collaborative relationship between the "federal funders" and "local doers." For example, if such a strategic planning process had been pursued in the Regalridge Square project, a more accurate, early analysis would have occurred of the potential for reaching the project's objectives. This might have precluded the accumulation of several million dollars of uncollectible debt with which the city of Fort Worth is now burdened.

2. Local officials must make an early, long-term public commitment of funds, political energy, and staff resources to revitalization projects. An uneven level of interest and support from city hall renders improbable the consistent and enthusiastic support of residents, property owners, and investors.

3. A critical ingredient in successful revitalization projects is public-private collaboration from the inception of project planning to the conclusion of implementation. Partnerships, however, should not be limited only to those providing dollars. All parties should be represented. Constant participation of public and private officials, project local leaders, and property owners rendered the downtown Fort Worth and Stockyard projects relatively successful. On the other hand, the difficulties of the Polytechnic project were caused in no small part by the distant, mandated shifts in objectives imposed by the funding agencies.

4. In neighborhood and commercial revitalization, leadership nurturing and training and ongoing recognition of the efforts of local leaders would measurably enhance the probability of project success. Federal revitalization policy for the past two decades has given sparse attention to preproject leadership development and ongoing support of leaders in project areas. Future policy guidelines, particularly for neighborhood and small commercial projects, should ensure early support for training and expert assistance to local leaders. The constant and active involvement of city, state, and local officials in their efforts should also be encouraged.

5. Downtown revitalization projects should be maintained and expanded with permanent, well-supported downtown associations. Downtown Fort Worth, Inc., as a successful example, has provided a host of critical security, maintenance, and promotional services that, in turn, have ensured the continued viability of the downtown projects.

6. Finally, wherever possible, the "strings" that are often attached to federal and foundation funding should be loosened and creative experimentation should be encouraged among those directly responsible for planning and implementing projects. Local participants in both the Polytechnic and Magnolia Street projects, for example, indicated that funding requirements often stifled local creativity. Fiscal control and accountability should be negotiated among all players during the planning process rather than simply imposed by those supplying funds.

In summary, revitalization in central cities is a complex, difficult process. Ongoing, collaborative involvement of both the public and private sectors in strategic planning and implementation is a critical key to success. Where large sums are required for large-scale physical development, such as in the Fort Worth downtown and Stockyard projects, public-private collaboration with well-capitalized, experienced private investors is imperative. Smaller-scale commercial or residential revitalization requires less concentrated private sector wealth, but it does require a core of supportive property owners led by an energetic and devoted local leader. Further, federal revitalization policy must be designed to ensure realization of these critical elements.

Notes

1. Material in this section is drawn from Downtown Fort Worth, Inc. (1992), and interviews conducted in summer 1993 with the following individuals: Doc Cornutt, chief financial officer, Woodbine Development Corporation, Dallas; Charles Cumby and Gene Foster, contract compliance specialists, Fort Worth Department of Housing and Human Services; Russell Lancaster, Fort Worth city councilman, 1981-1989; and Sharkey Stovall, former mayor of Fort Worth.

2. Material in this section is drawn from Fort Worth (1986), Mead and Miller (1989), and Weaver (1989).

3. Material in this section is drawn from interviews conducted in summer 1993 with Virginia Nell Webber, former Fort Worth mayor pro tem; Wayne Snyder, Fort Worth assistant planning director and industrial development coordinator; and Russell Lancaster, Fort Worth city councilman, 1981-1989.

4. Material in this section is drawn from interviews conducted in summer 1993 with Carol Becker, executive director of the North Fort Worth Business Association; Steve Murrin, partner in Billy Bob's Texas, White Elephant Enterprises, and Fort Worth Stockyards; Carolyn Snyder, Billy Bob's Texas, Fort Worth; and Jack Shannon, owner of Shannon Funeral Homes and president of the SARC.

5. Material in this section is drawn from Fort Worth (1989, 1990b) and from an interview conducted in summer 1993 with James Toal, former Fort Worth planning director.

6. Material in this section is drawn from interviews conducted in summer 1993 with Yvonne Coronado, volunteers coordinator, community and neighborhood drug offenses, Polytechnic Heights; Dorcas Gibson, president of the Polytechnic Heights Business Association; Reba Henry, chair of the Polytechnic Main Street Project board, 1986-1988; Terry Meza, project director of the Polytechnic Main Street Project, 1989-1992; Bryan Stone, executive director of Liberation Community, Polytechnic Heights; Tom Turner, director of economic and community development programs at Texas Wesleyan University; and Pam Walters, president of the Polytechnic Heights Homeowners Association.

7. Material in this section is drawn from interviews conducted in summer 1993 with Linda Beals, accountant with Liberation Communities, Inc., Regalridge Square; and Rosanne Briscoe, manager of the Regalridge Square apartments.

References

Barrett, K., & Green, R. (1992, February 18). The nutmeg doughnut: Why America's richest state has three of the country's poorest cities. *Financial World*, pp. 50-52.

Downtown Fort Worth, Inc. (1992). *Downtown fast facts*. Fort Worth: Author.

Fort Worth Chamber of Commerce. (1992). *Fort Worth metropolitan area 1992 statistical brief*. Fort Worth: Author.

Fort Worth, City of. (1960). *Annual budget*. Fort Worth: Author.

Fort Worth, City of. (1970). *Annual budget*. Fort Worth: Author.

Fort Worth, City of. (1977). *Inner city study: Social and economic analysis report*. Fort Worth: Office of Human Development.

Fort Worth, City of. (1980). *Annual budget*. Fort Worth: Author.

Fort Worth, City of. (1986). *Research and information, 1980-1985*. Fort Worth: Department of Planning and Growth Management, Economic Development and Neighborhood Revitalization Division.

Fort Worth, City of. (1989). *History of the Magnolia Avenue Project*. Fort Worth: Planning Department.

Fort Worth, City of. (1990a). *Annual budget*. Fort Worth: Author.

Fort Worth, City of. (1990b). *Medical district: A diversified business community—a development area profile*. Fort Worth: Planning Department.

Fort Worth, City of. (1991a). *Annual budget*. Fort Worth: Author.

Fort Worth, City of. (1991b). *Drug-related offenses*. Fort Worth: Department of Public Safety.

Lawrence Halprin & Associates. (1971). *CBS sector plan*. San Francisco: Author.

Mead, S., & Miller, C. J. (1989). Texas statutory tools for joint economic development: Public improvement district assessment and increment financing. In S. M. Wyman & R. R. Weaver (Eds.), *Texas economic development in transition: Opportunities for public-private collaboration*. Arlington: University of Texas.

North Central Texas Council of Governments, Regional Data Center. (1992). *1990 census summary, tape file 3A*. Arlington: Author.

Texas Employment Commission. (1992). *Arlington-Fort Worth PMSA employment statistics*. Fort Worth: Author.

Tyler, P. (1983, July 7). Bass' vision for downtown. *Fort Worth Star-Telegram*, p. C1.

U.S. Bureau of the Census. (various years). *Summary of population and housing*. Washington, DC: Government Printing Office.

Weaver, R. R. (1989). *Local economic development in Texas*. Arlington: University of Texas, Institute of Urban Studies.

6

Rebuilding Downtown

A Case Study of Minneapolis

ALEX SCHWARTZ

Festival marketplaces, convention centers, hotels, stadia, waterfront development, office development, mixed-use development, luxury housing, aquariums—these are some of the most popular types of urban development projects undertaken in U.S. cities during the past two decades. Few cities have pursued identical urban redevelopment strategies or constructed the same combination of redevelopment projects. Nevertheless, these types of projects constitute a virtual palette of redevelopment formats. Cities seem to select project ideas from this palette, applying them in various combinations and densities.

One city that has pursued almost every conceivable type of redevelopment project is Minneapolis. In little more than a decade, the city has built two sports stadia, one convention center, six hotels, four high-end shopping malls, and more than 13 million square feet of office space. The city has also pursued several riverfront development projects, including two festival marketplaces and a high-technology complex. The only element from the redevelopment palette still missing is an aquarium, and even that is under consideration. Most recently, Minneapolis has also embarked on an ambitious neighborhood redevelopment program in which revenue generated from downtown projects is channeled to residential neighborhoods for housing and a variety of other purposes.

Minneapolis's experience with urban redevelopment exemplifies both the possibilities and the limitations of a city's capacity to improve its economy, built environment, and neighborhood conditions. In this chapter, I examine redevelopment activity in Minneapolis since the 1950s, especially since 1980. I review the number and type of downtown and other redevelopment projects built during this period, focusing on the roles of the chief organizations involved in the development process and the financial mechanisms used to fund these projects. Downtown development is my central concern; I do not address Minneapolis's long history of neighborhood redevelopment, except in relation to downtown development. To place the analysis of the city's redevelopment in context, I begin with an overview of Minneapolis's historical development, demographic profile, economic base, and governmental structure.

City Context

The city of Minneapolis was first settled in 1848 by the St. Anthony Falls on the Mississippi River, just beyond the navigable limits of the river's northern reach. The city was established 20 years after the founding of St. Paul, located 10 miles downriver. These "Twin Cities" now constitute the center of a seven-county metropolitan region with a population in 1990 of 2.46 million. Whereas St. Paul was originally settled as a trading center on the Mississippi, Minneapolis relied on the river to power its lumber and flour mills. From the mid-nineteenth century until the first decades of the twentieth century, Minneapolis was known as Mill City. Milling activity declined sharply once regional timber supplies were depleted in the 1910s, and various transportation improvements, particularly the opening of the Panama Canal, reduced Minneapolis's competitive position as a flour-processing center by the 1930s. Today, there are few remaining vestiges of the city's previous mill-based economy, although several of the metropolitan area's most prominent companies, including Pillsbury and General Foods, originated as flour mills in the late nineteenth century. A wide variety of manufacturing industries came to replace milling as the mainstay of the city's economy. Following World War II, the city became a center for computer and other high-technology industries and for corporate headquarters in general (Abler, Adams, & Borchert, 1976).

POPULATION AND INCOME TRENDS

Minneapolis's population stabilized in the 1980s, the first decade since the 1950s when it did not decrease significantly. In 1990, the city had 368,000 residents, down 29% from its 1950 total of 522,000. Most of the city's population decline occurred before 1980; the population fell by less than 1% between 1980 and 1990.

The white population is the only racial group that has declined in numbers during the postwar period. It decreased by 11% in the 1980s, compared with 20% in the 1970s and 13% in the 1960s. In relative terms, the white population constituted 78% of the city's population in 1990, down from 87% in 1980 and 94% in 1970. The African American population, in contrast, has grown rapidly throughout the postwar period, although it remains small compared with that of other northern cities. By 1990 it made up 13% of the city's population, up from 8% in 1980 and 4% in 1970. The Native American population has also increased significantly, rising from 1.3% of the total population in 1970 to 2.4% in 1980, to 3.3% in 1990. The fastest rate of population growth has been posted by the Asian/Pacific Islander/other category. This group has increased from less than 1% of the total population in 1970 to more than 5% in 1990. The city's Hispanic population remains quite small. Although the Hispanic population increased by 53% in the 1970s and by 38% in the 1980s, its share of the city's total population has risen from 0.9% in 1970 to just 2.1% in 1990.

Minneapolis is one of the more affluent U.S. cities. Personal income tends to be well above the national average. Per capita income, adjusted for inflation, has increased from $9,495 in 1969 to $10,957 in 1979, to $11,520 in 1987 (Ashwood, 1991). Median family income, in contrast, has remained flat after inflation is taken into account, going from $25,632 in 1970 to $23,714 in 1980, to $25,243 in 1990.[1]

LOCAL ECONOMY

Minneapolis has long enjoyed a strong, diversified economy. As of 1990, the city's payroll employment totaled 286,000, an increase of 13,300 jobs since 1980 (Minneapolis, 1992, p. 51). The city of Minneapolis accounts for 39% of total employment in the surrounding metropolitan area (St. Paul accounts for an additional 26%). Unemployment has remained quite low in Minneapolis during the 1980s and 1990s, exceeding 5% only in 1982 (6.6%), 1983 (6.9%), and 1984 (5.1%) (Minneapolis, 1992).

Minneapolis is the leading corporate and financial center in the upper Midwest. Its economy has shifted from manufacturing, agricultural processing, and trade to corporate headquarters and producer services—financial, business, professional, and communications services and other information-intensive industries (Maki, 1990). This transformation, typical of several major U.S. cities, is best described by employment trends. In the 1980-1988 period,[2] the city of Minneapolis gained 12,429 nonagricultural jobs (Table 6.1). This 5% net increase reflected a loss of 14,838 jobs in construction; manufacturing; transportation, communications, and utilities; and wholesale trade. This was counterbalanced by an increase of 27,267 jobs in retail trade; finance, insurance, and real estate; services; and government. Goods-producing and goods-distributing industries lost employment, but service-producing industries and government more than made up for this loss. In relative terms, the goods-producing and goods-distributing sector's share of total nonfarm employment decreased from 34.5% to 27.8%, and the service-producing sector and government's share increased from 65.5% to 72.2%.

The ascendancy of producer services and other information-intensive industries is illustrated by the industries that posted the largest employment increase in Minneapolis in the 1980-1988 period (Table 6.2). Three producer services—security and commodity brokers, business services, and legal services—ranked among the city's top four job producers, yielding a total increase of 10,737. Printing and publishing was the only manufacturing industry to rank among the city's top employment growth leaders, and it is closely tied to several document-churning producer services, including banking, securities, advertising, insurance, and law. Other growth leaders included health and social services and eating and drinking places. On the bottom of the list is government—federal, state, and local.

Minneapolis, as noted earlier, is one of the nation's largest headquarters centers. As of 1982, Minneapolis was second only to Boston in the number of corporations per capita (Brandl & Brooks, 1982). The largest corporations with headquarters in the city of Minneapolis include Pillsbury, Honeywell, International Multifoods, and Bemis. Corporate Minneapolis has long been distinguished by its local origins and local loyalty. Most of the companies in Minneapolis are either headquartered

TABLE 6.1 Nonagricultural Employment, Minneapolis, 1980-1988

Industry	1980 Total	%	1988 Total	%	Change Total	%
Construction	7,554	2.8	6,321	2.2	-1,233	-16.3
Manufacturing	49,612	18.2	40,336	14.2	-9,276	-18.7
Transportation, communications, utilities	15,299	5.6	14,934	5.3	-365	-2.4
Wholesale trade	21,358	7.9	17,394	6.1	-3,964	-18.6
Retail trade	40,499	14.9	41,471	14.6	972	2.4
Finance, insurance, real estate	26,976	9.9	32,098	11.3	5,122	19.0
Services	69,510	25.6	86,186	30.3	16,676	24.0
Government	41,218	15.2	45,715	16.1	4,497	10.9
Total	272,026	100.0	284,455	100.0	12,429	4.6

SOURCE: City of Minneapolis (1990).

there or founded there (Ouchi, 1984, p. 189). Almost half of the manufacturing workers in Minneapolis are employed by local companies—the highest ratio in the nation (Ouchi, 1984, p. 189). In 1982, 11 of the city's 12 largest private employers were locally founded and remained headquartered there (Brandl & Brooks, 1982). Local loyalty is manifest in the active involvement of Minneapolis companies in civic affairs. Business leaders instigated the city's first major downtown and neighborhood development projects, and established one of the nation's most ambitious corporate philanthropy programs, in which companies dedicate 5% of their pretax earnings to charity (Ashwood, 1991; Fraser & Hively, 1987; Galaskiewicz, 1985; Ouchi, 1984).

The commitment of Minneapolis companies to civic affairs has come under mounting pressure since the mid-1980s. Increasing global competition, foreign takeovers, debt-financed acquisitions, industry shakeouts, recession, and the retirement of senior executives have forced many companies to scale back their civic and philanthropic contributions to the city. Pillsbury, a pillar of the Minneapolis corporate establishment, for example, was acquired by the British conglomerate Grand Metropolitan in 1986. The IDS Corporation became a subsidiary of American Express earlier in the 1980s. The metropolitan area's major banks suffered large losses during the farming crisis of the 1980s, and

TABLE 6.2 Industries With Most Employment Growth, Minneapolis, 1980-1988

Industry	Employment Growth	
	Total	%
Security and commodity brokers	3,856	110.2
Health services	3,511	16.2
Business services	3,486	22.1
Legal services	3,395	98.9
Eating and drinking places	2,923	23.3
Printing and publishing	2,346	26.2
Social services	2,138	10.9
Local government	1,808	8.9
Federal government	1,375	28.5
State government	1,315	8.2

SOURCE: City of Minneapolis (1990).

its computer companies, among them Honeywell and Control Data, have suffered from industrial restructuring and reduced defense procurement (Ashwood, 1991; Worthy, 1987). Finally, many of the city's key executives who had been most active in civic affairs have retired, often replaced by people, usually from out of state, whose long-term commitments to the city are not known. These changes in the corporate community have sparked speculation about the future of business involvement in the public sphere. For example, the new generation of corporate leaders is regarded as less likely than their predecessors to attend civic meetings, and instead delegate this activity to officials with less authority to commit corporate resources to public projects (Worthy, 1987). In terms of corporate philanthropy, however, there is no evidence to show that Minneapolis companies are less generous than before (Anderson, 1990; *Focus,* 1992).

LOCAL GOVERNMENT STRUCTURE

The mayor and the 13 city council members who govern Minneapolis are elected to 4-year terms, with council members representing individual wards. The city council wields the most power within city government. It is responsible for setting citywide policies, approving the city budget, passing all city ordinances, and overseeing all city agencies except the Parks and Library Departments, which are managed by independent boards. City council members also serve as the Board of Directors for the Minneapolis Community Development Agency (MCDA), a quasi-independent authority responsible for housing and economic development. The mayor proposes annual budgets to the council, presents policy proposals, and has the authority to veto council decisions. The mayor also hires and fires department heads and appoints individuals to boards and committees.

Several government functions are outside the purview of the mayor and city council. The Parks and Library Departments are managed by independent boards, although they are funded by the city government. The public schools are operated and financed by an independent school district with its own taxing authority. Health and social services, solid waste management, roads, courts, and other services are provided by Hennepin County. Approximately one-fourth of the property tax in Minneapolis is levied by the city government, one-fourth by the county, and half by the school district.

INTERGOVERNMENTAL RELATIONSHIPS

Unlike all other major metropolitan areas, the Twin Cities region is subject to several forms of metropolitan government. Most unique is the region's Fiscal Disparities program, in which a portion of the region's commercial-industrial tax base is redistributed to individual municipalities according to their fiscal capacities.

Fiscal Disparities

The Fiscal Disparities program was initiated by the Minnesota State Legislature in 1971 to reduce inequities among municipalities in the seven-county Twin Cities region, to diminish intraregional competition for economic development, and to discourage urban sprawl (Baker, Hinze, & Manzi, 1991; Goetz, 1992; Lassar, 1991; Metropolitan Council, 1991). Litigation, however, delayed the start of the program until 1975. The Fiscal Disparities program is structured so that 40% of the growth in each municipality's commercial-industrial (C-I) tax base since 1971 is pooled into a single fund, which is then redistributed to municipalities according to their populations and fiscal capacities. New development and inflation have steadily increased the value of the region's post-1971 C-I tax base. For taxes payable in 1991, the Fiscal

Disparities program redistributed $290.5 million, or 30.8% of the region's total C-I tax base. For taxes payable in 1980, the program controlled less than 15% of the regional C-I tax base (Metropolitan Council, 1991).

As noted, Fiscal Disparities funds are distributed to municipalities according to their populations and fiscal capacities, that is, the equalized per capita market value of all residential and nonresidential property.[3] Communities with less real property value per capita receive more Fiscal Disparities funds than do communities with greater valuation per capita. More than three-fourths of the 188 cities and townships in the metropolitan area receive more Fiscal Disparities funds than they contribute, and 31 contribute more than they receive (Metropolitan Council, 1991).

The Fiscal Disparities program has significantly reduced fiscal inequities among communities in the Twin Cities region. Without the program, the ratio of the highest per capita tax base to the lowest would be 22 to 1; with the program, the ratio is just 4 to 1 (Metropolitan Council, 1991). The effect of the program on municipal tax rates, however, is much less pronounced. According to a recent analysis by the Minnesota House of Representatives Research Department, elimination of the Fiscal Disparities program would, with only a few exceptions, cause only trivial changes in local tax rates. This is because state aid to schools and municipalities partially offsets the distribution of Fiscal Disparities revenue (Baker et al., 1991).

Hennepin County

Minneapolis is the largest city in and the county seat of Hennepin County (St. Paul is in Ramsey County). As of 1990, the city accounted for 35% of the county's population of 1.03 million, and 39% of its total employment of 732,452. The county is responsible for social services, health services, solid waste management, roads, and the courts. The county and the city have had a testy relationship over the years. The city, according to several observers, tends to regard the county government as unsympathetic to its needs, more than occasionally treating the city as a dumping ground for prisons and other controversial facilities. The county, on the other hand, thinks the city should consult with it more on major projects. As will be discussed later, the county and city are especially at odds over the use of tax increment financing, the primary funding source for the city's redevelopment.

Downtown Development

Until the 1980s, Minneapolis's downtown buildings seemed to belie the city's importance as one of the nation's largest corporate centers. The city's first large office building, the Foshay Tower, was completed in 1929. It dominated the skyline for 43 years, until 1972, when the 57-story IDS building was constructed and, according to one local architect, was so out of scale it made the rest of Minneapolis look like a "toy town" (Hauser, 1990). Although the decade of the 1970s was then perceived as a building boom for downtown Minneapolis, the construction of this period was nothing compared with the 1980s. A total of 27 new downtown projects were completed in the 1970s, as opposed to more than 70 new projects in the 1980s and at least 15 more in the early 1990s. In addition to new construction, more than 80 major rehabilitation projects were undertaken in the 1980s. In addition to downtown, the city's long-neglected riverfront has also undergone redevelopment, although with mixed results. Another area transformed is the city's traditional warehouse district, where warehouse and manufacturing facilities have been converted, with essentially no public assistance, into artists' studios, galleries, theaters, restaurants, and bars (Roe & Rucker, 1991).

In this section I review the history of the city's downtown development, focusing on major projects and the key organizations involved. In subsequent sections I examine development subsidies and development impacts.

DOWNTOWN REDEVELOPMENT
IN THE 1950s AND 1960s

Gateway

Minneapolis's first downtown redevelopment project began in the 1950s under the federal urban renewal program. In 1957, a coalition of business executives and government officials secured urban renewal funds to redevelop approximately one-third of the downtown area. Known as the Gateway area, this was the oldest section of downtown and had become the city's skid row, containing rooming houses, taverns, cheap restaurants, and pawn shops.

The Gateway renewal project encompassed 76 acres, within which 186 buildings were demolished, displacing 2,400 people (mostly single men) and 450 businesses. A few of the destroyed buildings were of architectural significance and would probably have been spared had the Gateway project begun a few years later, when historic preservation was of greater concern (Martin & Goddard, 1989). The total cost of the Gateway project was $25 million, 75% of which was furnished by the federal government, 25% by the city. As with urban renewal projects throughout the nation, Gateway redevelopment was subsidized through a "land write-down," in which a portion of the cost of land assembly and improvement was assumed by the public sector.

More than $60 million in new construction in the Gateway area was either committed or completed by early 1963. New structures built in the Gateway area included the Minneapolis Public Library, the state Employment Security Office, a new federal courthouse, a few office buildings, a hotel, and two apartment buildings (Martin & Goddard, 1989, p. 64). Only a quarter of the Gateway project had been completed by 1964. Development then languished for want of interested developers until the late 1970s. The only substantial new development in the intervening years involved two high-rise apartment buildings for senior citizens and an office building for the Federal Reserve Bank. Much of the remaining cleared land was used for parking lots, which generated substantial income for the city. The area's total private development, consisting mostly of single-use office buildings of scant architectural value, amounted to about $110 million (Ashwood, 1991; Martin & Goddard, 1989; Millet, 1989).

The Gateway project shaped subsequent redevelopment efforts in Minneapolis in at least two ways. First, it provided vacant, assembled, and improved land for future projects (Martin & Goddard, 1989). Second, it established the precedent of the land write-down as a key development subsidy, which became the basis of much of the city's redevelopment supported by tax increment financing (TIF) in the 1970s, 1980s, and 1990s.

Nicollet Mall and the Skyway System

In 1955, downtown business leaders formed the Downtown Council (DTC). This organization reflected the concern of downtown retailers, bankers, newspaper publishers, and other business executives over the viability of downtown Minneapolis. This concern was based partly on the lack of any new office construction since 1929. It was heightened when General Mills relocated its corporate headquarters in 1955 from downtown Minneapolis to the suburbs, and when the metropolitan region's first suburban shopping mall opened in 1956 (Aschman, 1971; Ashwood, 1991). In its first years, the DTC commissioned an impact study of the then-fledgling interstate highway system, and it successfully lobbied the city to expand the Planning Department under new leadership (Aschman, 1971; Ashwood, 1991).

The DTC worked with, and funded, the newly revived City Planning Department on a "comprehensive downtown plan." Never completed after a draft report was released in early 1960, and largely ignored by the city council and other city agencies, the plan nevertheless served as a blueprint for subsequent redevelopment activity. It articulated five major concepts for downtown: (a) a bypass ring of highways around the downtown, (b) parking terminals at highway exits, (c) clustering of different types of development within separate zones, (d) pedestrian routes and specialized street use, and (e) creation of super blocks. The first four of these concepts have been realized in publicly and privately initiated redevelopment projects (Ashwood, 1991; Goldfield, 1976).

The first downtown developments to reflect the priorities of the comprehensive plan produced by the City Planning Department in close collaboration with the Downtown Council were the Nicollet Mall and the skyway system. The first involved the conversion of downtown's primary shopping street into a limited-access transit and pedestrian mall. The second is a network of elevated, glass-enclosed walkways, or "skyways," connecting office buildings, banks, department stores, and parking garages.

The Nicollet Mall is Minneapolis's second and still best-known downtown redevelopment project. After commissioning a consultant in 1959 to establish objectives and review alternatives for Nicollet Avenue, the DTC decided to turn Nicollet Avenue into a "transit mall," with the street open only to buses, taxis, and emergency vehicles. It also decided to proceed with the consultant's suggestion to establish a system of skyways connecting parking garages with the second floors of buildings throughout the mall area. The scale and cost of the plan required the DTC to work with local, state, and federal government agencies to secure financing and regulatory approval. The total cost of the mall, not including the skyway system, amounted to $3.7 million, primarily for utility and other underground infrastructure improvements.

Working with several government authorities, the DTC devised a special-benefit assessment to finance the bulk of the project, by which downtown property owners paid most of the costs, with properties at the core of the mall bearing the greatest financial burden (Aschman, 1971; Ashwood, 1991; Brandl & Brooks, 1982; Goldfield, 1976). Upon its completion in 1967, 10 years after the DTC first sought ways to strengthen the downtown area, the Nicollet Mall received national attention as one of the most innovative and successful downtown revitalization projects in the country (Breckenfeld, 1976; Goldfield, 1976; Mitchell, 1974; Wiedenhoeft, 1975).

The skyway system, like the Nicollet Mall, was promoted by the DTC as a way to help fortify downtown Minneapolis. The construction of enclosed, weather-proof pedestrian walkways was recommended in consultant Barton-Aschman's 1960 report to the DTC on downtown revitalization, and even earlier by several local businessmen (Ashwood, 1991). The first skyways were built in 1962, five years before the Nicollet Mall. As of 1992, Minneapolis had 46 skyways (and two tunnels) connecting 36 city blocks. Calgary is the only city with a more extensive system (John J. Labosky, president, Minneapolis DTC, interview, June 1992; see also Ashwood, 1991; Lassar 1989). Unlike St. Paul, where the skyway system is publicly owned, all of Minneapolis's skyways are under private ownership and management. The Minneapolis skyways are also an important component of the downtown retail market, offering almost 370,000 square feet of second-story space with net rents between $20 and $30 per square foot (MCDA, 1991b, p. 14).

DOWNTOWN DEVELOPMENT IN THE 1970s

Although the pace and magnitude of downtown development reached record-breaking levels in the 1980s and early 1990s, several major downtown projects were initiated in the 1970s. These include the Loring Park Development District, the IDS Center, and the Hennepin County Government Center.

The Loring Park Development District encompasses a nine-block residential area immediately south of downtown. The city originally envisioned Loring Park as an urban renewal project, but federal funds were no longer available by the time the project was under way. Instead, the city funded the project by establishing a tax increment district in 1972. Through tax increment financing, the city has invested approximately $37 million for land acquisition and clearance, relocation assistance for residents and businesses, and construction of public improvements.

The Loring Park Development District involved several major projects, some of which were not completed until the early 1980s. These included several large housing developments, with about one-fourth of the total units reserved for low- and moderate-income residents; a Hyatt Regency Hotel and merchandise mart; Orchestra Hall;[4] and campuses for two educational institutions (Ashwood, 1991; Wiedenhoeft, 1975). In addition, the Loring Park Development District financed the four-block extension of the Nicollet Mall toward Loring Park.

Although the Loring Park District took much longer to develop than was originally expected, it is now widely considered to be a financial and planning success. Before the tax increment district was established in 1972, the Loring Park area generated less than $400,000 a year in real estate taxes. By 1990, it yielded more than $5 million annually (Ashwood, 1991; Minneapolis, 1992).

The IDS Center was by far the most dramatic addition to the Minneapolis skyline in the 1970s, and remains the city's tallest building today. Designed by the celebrated New York architects Philip Johnson and John Burgee and Minneapolis architect Edward Baker, the IDS Center includes a 51-story office tower, a 19-story hotel and bank building, and another 8-story office building. Connecting these structures is the Crystal Court, a two-level glass atrium containing a shopping center and restaurant court. The 2.5 million-square-foot development cost $42 million to construct and was financed with no government assistance (Ashwood, 1991; Wiedenhoeft, 1975).

The Hennepin County Government Center houses courts and administrative offices for Hennepin County. Its twin towers were completed in 1973 at a cost of $41.8 million. Besides its prominence in the downtown skyline, the Government Center is significant for its location several blocks east of Nicollet Mall. It was one of the first office buildings to be built in the area of downtown designated in the city's original downtown plan for office development (Ashwood, 1991; Wiedenhoeft, 1975).

DOWNTOWN DEVELOPMENT SINCE 1980

Minneapolis, like many other cities, experienced a massive building boom in the 1980s. As noted earlier, the decade saw construction of office buildings, shopping centers, housing, hotels, sports facilities, a convention center, parking garages, and restoration of major theaters. More than 85 new projects were built downtown from 1980 through 1991, at a cost exceeding $3.4 billion. Of these, 30 benefited from tax increment financing (TIF) and 10 received other public subsidies (Ashwood, 1991; MCDA, 1984, 1986, 1987, 1991a, 1991b).

Office Development

Office buildings constitute the single largest component of Minneapolis's downtown redevelopment in the 1980s. Almost 4 in 10 of the new downtown projects of the 1980s and early 1990s were either single-purpose office buildings or mixed-use developments with extensive amounts of office space. In total, the city has absorbed more than 750,000 square feet of office space annually since 1980, the equivalent of one 40-story office building per year.

The largest new office buildings constructed since 1980 include the following:

- the Pillsbury Center (1.4 million square feet), completed without government assistance in 1981 at a cost of $80 million
- the Multifoods Tower, the final part of the City Center project, completed in 1982 without government assistance at a cost of $150 million
- the Piper Jaffrey Tower (700,000 square feet), completed at a privately financed cost of $95 million
- the Norwest Center (1.1 million square feet), designed by Cesar Pelli and completed in 1988 for $111 million, all privately financed
- the 150 South 5th Street building (650,000 square feet), completed in 1988 without public subsidy at a cost of $100 million
- the IDS Data Center (525,000 square feet), completed in 1992 with tax increment financing
- the LaSalle Plaza (1.1 million square feet), a mixed-use project supported with tax increment financing, includes 747,000 square feet of office space completed in 1991
- the Dain Bosworth Plaza (795,000 square feet), completed in 1991 with tax increment financing, at a cost of $175 million, includes 600,000 square feet of office space as well as 120,000 square feet of retail space anchored by a Neiman Marcus department store
- the First Bank Tower (1.4 million square feet), privately financed at a cost of $150 million, completed in 1992
- the International Center III (607,000 square feet), a facility leased by AT&T, completed in 1991 at a cost of $100 million with no public subsidy

Fewer than half of the new office structures received TIF or other subsidies. When the city does provide financial assistance to office development, it usually does so to achieve specific ancillary objectives. For example, the $175 million Dain Bosworth Plaza received $1 million in tax increment financing to support the Neiman Marcus department store at the tower's base; by including an office tower and its large tax base within the TIF district, the city could better ensure the financial success of the district than if only a freestanding department store had been built. Similarly, the city provided $19.5 million to the mixed-use LaSalle Plaza and its 747,000 square feet of office space to help stimulate development of Hennepin Avenue (James E. Moore, director of economic development, MCDA, interview, June 1992).

Retail Development

Retailing has long been a central concern of downtown planners. The effort to remake Nicollet Avenue into the Nicollet Mall was instigated in the 1950s after the opening of the first suburban shopping mall. Suburban competition has intensified ever since. The downtown and the rest of the city have accounted for a dwindling share of the metropolitan region's retail sales (Ashwood, 1991). Three major downtown department stores have closed since 1980: Powers in 1985, J.C. Penney in 1986, and Carson Pirie Scott in 1993. Given increased suburban competition, the proliferation of suburban shopping centers anchored by major department stores, the Minneapolis Community Development Agency (MCDA) and the Downtown Council have attempted to reposition the downtown as a unique, upscale shopping district within the greater metropolitan region. In addition to Dayton's and other department stores that operate at multiple suburban locations as well as downtown, the city has attempted to bring in distinctive high-end stores that are not represented elsewhere in the region. To this end, the city established several TIF districts to finance the construction of downtown shopping centers anchored by Saks Fifth Avenue, Neiman Marcus, and other luxury retailers.

The city's first major downtown retail project of the 1980s was the City Center, a mixed-use development featuring a 52-story office tower, a hotel, and a suburban-style mall with 90 stores and restaurants anchored by Carson Pirie Scott. The $103.5 million, 1 million-square-foot project was completed in 1982 with the assistance of tax increment financing. Although maligned for its fortresslike design and its aloofness from the street (Lu, 1984), the City Center has been a tremendous success in terms of tax generation. The development yields almost $10 million each year in tax increment revenue, accounting for more than 17% of the city's total tax increment revenue (Minneapolis, 1992). The Carson Pirie Scott store

closed in January 1993, but its space was subsequently taken over by
Filene's Basement, Montgomery Ward, and other national retailers.
Other major retail developments came on line during the late 1980s
and early 1990s. They include the Conservatory, a $75-million anchor-
less complex completed in 1987, consisting of expensive boutiques
restaurants, and offices, which is now in financial trouble; Gaviidae
Common, an $84 million shopping center, completed in 1989 and
anchored by Saks Fifth Avenue; and the $175 million Dain/Neiman
Marcus project, completed in 1990, which combines a Neiman Marcus
store with a large office building. All three of these retail projects received
tax increment financing. Gaviidae Common and the Dain/Neiman
Marcus developments are both situated within the City Center tax
increment district.

In addition to these new retail centers, the city also has invested $21.8
million in the early 1990s to renovate the Nicollet Mall, which had
physically deteriorated since its opening in 1967. To finance the renovation, downtown property owners were assessed for 85% of the total
cost, with assessments based on the property's proximity to the mall.
The city paid for the remaining 15% of the project's cost from its
operating budget.

Housing

Before the 1980s, downtown Minneapolis was mostly nonresidential
Although tens of thousands of people worked and shopped downtown
relatively few people lived there. Several major housing developments
were completed in the 1980s and early 1990s, increasing the area's
population by 58%, from 15,235 in 1980 to 24,280 in 1990 (J. J
Labosky, interview, June 1992). A total of 21 new projects with more
than 3,700 units were constructed downtown from 1980 through 1991
at a total cost of more than $330.5 million. All but three of these
developments received TIF or other public subsidies (Garner, 1991
MCDA, 1991a, p. 17).

Hotels

Six new hotels were constructed downtown during the 1980s, and
four others underwent major renovations.[5] In addition, an 816-room
Hilton Convention Hotel opened in November 1992, increasing the

supply of downtown hotel rooms by 20% (Wallace, 1992). Only one of the new hotels was built without government subsidy: a $14 million, 200-room Luxeford Suites Hotel. The others benefited from TIF and, in two cases, federal Urban Development Action Grants (UDAGs). The new Hilton Convention Hotel was built at a cost of $144.8 million, of which $85.5 million was funded with TIF and $3.8 million from UDAG. The MCDA has retained a 50% ownership stake in the Hilton Convention Hotel because of difficulty in finding private investors. The city decided to proceed with the project because it considers the hotel essential to the success of the new convention center (J. E. Moore, interview, June 1992).

Sports Facilities

The Hubert H. Humphrey Metrodome, home stadium for the Minnesota Twins and the Minnesota Vikings, opened in 1982. Its $55 million cost was financed by bonds backed by operating revenue and city excise taxes on hotel rooms and liquor sales. According to the executive director of the Metropolitan Sports Facilities Commission, the Metrodome is the only professional sports facility in the nation that "pays all of its debt service, operating costs, and capital improvements out of revenues" (quoted in Klobuchar, 1992). Since 1983, annual attendance has averaged 3.1 million (Klobuchar, 1992).

The Target Center, completed in 1990 at a cost of $94 million, is the arena for the Minnesota Timberwolves, the city's National Basketball Association franchise. Construction was assisted with tax increment bonds and other loans totaling $23 million. When not used for basketball, the center holds rock concerts and other events. In 1991, the Target Center held 205 events, including 41 basketball games, with attendance averaging 15,000 per event (J. J. Labosky, interview, June 1992).

Convention Center

The Minneapolis Convention Center opened in October 1990 at a cost of $150 million. The 775,000-square-foot facility contains 300,000 square feet of exhibition space and can accommodate 30,000 people. The center includes two 100,000-square-foot halls as well as a 31,400-square-foot ballroom and 58 meeting rooms. The project was financed with tax-exempt revenue bonds and sales tax revenue. Approximately

25 to 30 conventions are booked per year at the facility (J. J. Labosky, interview, June 1992; Wallace, 1992).

Hennepin Avenue Theaters

Hennepin Avenue, located one block west of the Nicollet Mall, was downtown Minneapolis's entertainment center until the 1950s. However, according to the MCDA (1990b), "over the last few decades the once-thriving theater and restaurant district fell into severe decline, falling prey to adult theaters and bookstores, drug traffic, prostitution, and the highest crime rate in the downtown area" (p. 10). Since the late 1980s, however, the city has revived Hennepin Avenue's theaters. In 1988, the city purchased and renovated the 2,769-seat Orpheum Theatre for a total of $2.7 million. Built in 1921, the theater is now operated under private management for Broadway road shows, concerts, and other events. In 1990, the city restored Hennepin Avenue's ornate State Theatre as part of LaSalle Plaza, a $40 million mixed-use development supported with tax increment financing.

Parking Facilities

The provision of ample commuter parking has long been an integral part of downtown development planning. Parking garages, known in Minneapolis as parking ramps, have been built on the periphery of downtown to provide people with parking outside the downtown core. As of 1991, 10 city-owned parking ramps were in operation, providing 10,472 parking spaces. Five additional ramps with 8,435 spaces were then under construction. In addition, the downtown has six existing ground-level parking lots with 1,527 spaces. Of the 10 existing ramps, 9 were built in the 1980s. All of the ramps are linked to other downtown buildings by skyways or, in one case, by tunnel. All of the ramps under construction are also to be linked to the skyway system, protecting commuters from Minneapolis's frigid winter temperatures (MCDA, 1991b).

KEY ORGANIZATIONS IN REDEVELOPMENT

The lead agency responsible for downtown, riverfront, and neighborhood development since 1981 has been the Minneapolis Community Development Agency. Created in 1980 with the consolidation of several city agencies, the MCDA is the development arm of the city of Minneapolis. Like development authorities in other cities, the MCDA is an independent governmental entity whose staff is exempt from civil service rules and whose actions are not directly accountable to the electorate (Fainstein & Fainstein, 1991; Leitner & Garner, 1993). Unlike its counterparts in other cities, however, the MCDA's board of directors consists of the city council. The original board was composed of members appointed by the mayor and the city council, but the city council decided in 1986, in opposition to the mayor, to dismiss this board and replace it with the city council members. As a result of this change, the MCDA is probably less insulated from city politics than are similar authorities in other cities.

The MCDA is responsible for most of the city's commercial, industrial, and housing development. It was also the city's public housing authority until 1991. As of 1991, the agency had a staff of 205 and an annual budget of $400 million—almost as large as the municipal budget. The MCDA obtains funds from a number of sources. It administers several federal and state grant and loan programs (Community Development Block Grants, Small Business Administration and Minnesota Housing Finance Agency grants and loans), issues industrial and housing revenue bonds, is responsible for tax increment financing, and receives revenue from loan payments, land sales, lease payments, and parking fees (MCDA, 1990a).

The agency administers nearly 50 programs, including homeownership development programs, home improvement programs, rental housing development programs, and commercial and industrial development programs. It also sponsors tax increment financing, a citizen participation and technical assistance program, an arts economic development assistance program, and other programs. The MCDA has also served as a consultant to other U.S. and foreign cities, including Chicago, Toledo, and St. Petersburg (formerly Leningrad).

Other public and nonprofit organizations have also been involved in downtown redevelopment, notably the City Planning Department and the Downtown Council. The City Planning Department plays a subsidiary role in the development process. It works on site plans for specific projects, and it has produced, in conjunction with the Downtown Council, three long-term downtown plans (Minneapolis Planning Department and Downtown Council of Minneapolis, 1988). The Planning Department is occasionally at odds with the MCDA over the direction of downtown development. Planners have criticized the MCDA for making expedient choices, for being too willing to make deals without giving adequate consideration to their long-term significance. For example, planners believe the city should be more selective in choosing tenants for its projects and should try to target specific growth industries rather than accept the first available company.

The Downtown Council continues to be an important force in downtown development, although its influence is less dominant than in the past. As discussed earlier, the DTC was the catalyst behind the Nicollet Mall and the skyway system. It also provided financial and organizational support for the Loring Park Development District (Ashwood, 1991). The power of downtown business leaders, however, is not what is was when, it is often remarked, a handful of key executives would decide on major projects over lunch at the Minneapolis Club. Now, downtown businesses, represented by the Downtown Council, are one of several actors shaping city development, above all, the MCDA and the city council, its board of directors. Still, the Downtown Council continues to serve several downtown functions, and its effectiveness is said to have increased since a period of leadership turnover in the mid-1980s.

Development Subsidies

EXPENDITURES BY LOCATION, PURPOSE, AND REVENUE SOURCE

The city of Minneapolis spent $334.8 million to assist downtown development projects in the 1980-1990 period, and $144.0 million since 1986. Of the total downtown expenditures since 1980, 34% was for commercial development (retail, office, hotel), 34% for industrial development, 20% for mixed-use development, and 12% for housing. In the 1986-1990 period, commercial uses accounted for 53% of total downtown expenditures, mixed-use projects 24%, industrial 21%, and housing 2%. In terms of total citywide development expenditures since 1980, downtown claimed 57% of commercial subsidies, 32% of industrial subsidies, and 7% of housing subsidies (Table 6.3).

Downtown development projects accounted for 26% of the city's total development assistance of $1.30 billion in the 1980-1990 period, and 32% of the city's $456.8 million in development subsidies since 1986. Riverfront projects constituted 15% of total city assistance in

1980-1990, and 11% in 1986-1990. Neighborhood development projects took the lion's share of the city's total development assistance in the 1980s. Neighborhood projects represented 59% of total expenditures in 1980-1990 and 58% in 1986-1990. Although neighborhood projects continued to absorb the bulk of the city's development subsidies in the 1980s, their share decreased substantially from the second half of the 1970s, when neighborhoods commanded 84% of the total (MCDA, 1991a).

The city of Minneapolis draws from a variety of revenue sources to fund its development activity. From the federal government, it has used Section 8 housing subsidies, Community Development Block Grants (CDBGs), Urban Development Action Grants, and Urban Mass Transportation Administration (UMTA) grants. It has also offered low-interest industrial revenue bonds (IRBs) and housing revenue bonds (HRBs), whose interest payments were exempt from federal taxes until the 1986 Tax Reform Act went into effect. The single most important type of development funding originating from local sources is tax increment financing. Other local revenue sources include loan repayments, property sales, lease income, and parking fees. The state of Minnesota has provided very few direct or indirect subsidies for redevelopment and other economic development projects. For the entire 1980-1990 period, direct federal assistance (CDBGs, Section 8 subsidies, UDAGs, and UMTA grants) accounted for 37.2% of the city's total development expenditures. Tax-exempt IRBs and HRBs made up an additional 36.7% of total expenditures, TIF 25.7%, other local revenues 2.6%, and state aid less than 1% (see Table 6.4).

As in all U.S. cities, federal aid to Minneapolis diminished throughout the 1980s as the government eliminated the UDAG program and cut back CDBG, Section 8, and other programs. Average annual revenue from Washington declined from $71.1 million in the 1975-1980 period to $49.4 million in 1981 through 1985, to $25.6 million in 1986 through 1990 (Table 6.4). As a proportion of the city's development expenditures, direct federal revenue declined from 72% in 1975 through 1980 to 34% in 1981 through 1985, to 28% in 1986 through 1990. In addition to these federal cutbacks, the national Tax Reform Act of 1986 essentially eliminated the appeal of IRBs and HRBs by removing their tax-exempt status. Before the act went into effect, developers could save millions of dollars by borrowing at lower interest rates because investors did not have to pay federal income taxes on IRB and HRB interest payments. With the exemptions phased out, local governments could no

(text continued on page 188)

184

TABLE 6.3 MCDA Program Expenditures by Location and Purpose ($ thousands)

Expenditure Purpose	Total Expenditures				Average Annual Expenditures			
	1975-1980	1981-1985	1986-1990	1980-1990	1975-1980	1981-1985	1986-1990	1980-1990
Downtown								
housing	11,208.5	37,417.6	2,965.9	40,383.4	1,868.1	7,483.5	593.2	3,671.2
commercial	28,462.8	33,710.4	76,190.6	114,099.5	4,743.8	6,742.1	15,238.1	10,372.7
industrial	10,733.0	74,856.6	29,834.0	114,257.1	1,788.8	14,971.3	5,966.8	10,387.0
mixed	23,432.5	20,853.4	34,728.4	65,594.3	3,905.4	4,170.7	6,945.7	5,963.1
miscellaneous	0.0	109.4	328.8	438.2	0.0	21.9	65.8	39.8
total	73,836.8	166,947.3	144,047.8	334,772.4	12,306.1	33,389.5	28,809.6	30,433.9
Neighborhoods								
housing	407,912.6	249,557.0	124,848.6	430,567.7	67,985.4	49,911.4	24,969.7	39,142.5
commercial	11,879.6	21,154.5	10,976.8	35,570.7	1,979.9	4,230.9	2,195.4	3,233.7
industrial	42,250.6	114,586.2	72,951.0	196,190.9	7,041.8	22,917.2	14,590.2	17,835.5
mixed	14,798.8	22,460.7	27,036.9	59,658.2	2,466.5	4,492.1	5,407.4	5,423.5
miscellaneous	18,809.2	18,244.7	28,873.8	49,795.8	3,134.9	3,648.9	5,774.8	4,526.9
total	495,650.8	426,003.0	264,687.1	771,783.3	82,608.5	85,200.6	52,937.4	70,162.1
Riverfront								
housing	7,184.2	83,070.0	18,216.5	105,823.0	1,197.4	16,614.0	3,643.3	9,620.3
commercial	11,102.9	18,572.5	29,039.4	49,318.6	1,850.5	3,714.5	5,807.9	4,483.5
industrial	787.5	35,158.9	780.0	35,938.9	131.2	7,031.8	156.0	3,267.2
mixed	0.0	0.0	0.0	0.0	0.0	0.0	0.0	0.0
miscellaneous	0.0	44.0	1.9	45.9	0.0	8.8	0.4	4.2
total	19,074.6	136,845.4	48,037.8	191,126.3	3,179.1	27,369.1	9,607.6	17,375.1

City total

housing	426,305.3	370,044.5	146,031.0	576,774.1	71,050.9	74,008.9	29,206.2	52,434.0
commercial	51,445.4	73,437.3	116,206.8	198,988.7	8,574.2	14,687.5	23,241.4	18,089.9
industrial	53,771.1	224,601.7	103,565.0	346,386.8	8,961.8	44,920.3	20,713.0	31,489.7
mixed	38,231.4	43,314.1	61,765.3	125,252.5	6,371.9	8,662.8	12,353.1	11,386.6
miscellaneous	18,809.2	18,398.1	29,204.5	50,279.9	3,314.9	3,679.6	5,840.9	4,570.9
total	588,562.2	729,795.8	456,772.7	1,297,682.0	98,093.7	145,959.2	91,354.5	117,971.1

SOURCE: MCDA (1991a).

TABLE 6.4 MCDA Program Expenditures by Revenue Source ($ thousands)

Revenue Source	Total Expenditures				Average Annual Expenditures			
	1975-1980	1981-1985	1986-1990	1980-1990	1975-1980	1981-1985	1986-1990	1980-1990
Downtown								
tax increment financing	48,145.3	45,194.4	98,462.0	154,117.5	8,024.2	9,038.9	19,692.4	14,010.7
UDAGs	0.0	681.1	10,606.2	11,287.3	0.0	136.2	2,121.2	1,026.1
IRBs	10,733.0	74,851.2	29,479.5	113,897.0	1,788.8	14,970.2	5,895.9	10,354.3
HRBs	11,208.5	36,260.6	0.0	36,260.6	1,868.1	7,252.1	0.0	3,296.4
CDBGs	0.0	0.0	0.0	0.0	0.0	0.0	0.0	0.0
Section 8	0.0	0.0	0.0	0.0	0.0	0.0	0.0	0.0
other	73,836.8	166,947.3	144,047.8	334,772.4	12,306.1	33,389.5	28,809.6	30,433.9
total	3,750.0	9,960.0	5,500.0	19,210.0	625.0	1,992.0	1,100.0	1,746.4
Neighborhoods								
tax increment financing	28,421.7	47,991.3	49,299.1	103,616.6	4,737.0	9,598.3	9,859.8	9,419.7
UDAGs	0.0	4,387.3	23,811.1	28,198.4	0.0	877.5	4,762.2	2,563.5
IRBs	26,633.0	104,552.1	58,634.6	167,890.9	4,438.8	20,910.4	11,726.9	15,262.8
HRBs	23,828.9	41,803.7	10,373.0	53,089.8	3,971.5	8,360.7	2,074.6	4,826.3
CDBGs	98,155.2	78,293.0	63,062.7	164,325.3	16,359.2	15,658.6	12,612.5	14,938.7
Section 8	297,560.4	146,126.9	59,506.7	235,208.1	49,593.4	29,225.4	11,901.3	21,382.6
other	495,650.8	426,003.0	264,687.1	771,783.3	82,608.5	85,200.6	52,937.4	70,162.1
total	21,051.6	2,848.7	0.0	19,454.2	3,508.6	569.7	0.0	1,768.6

Riverfront

tax increment financing	9,932.9	22,192.0	47,255.9	75,691.1	1,655.5	4,438.4	9,451.2	6,881.0
UMTA	0.0	44.0	1.9	45.9	0.0	8.8	0.4	4.2
UDAGs	6,049.4	4,544.8	0.0	4,544.8	1,008.2	909.0	0.0	413.2
IRBs	787.5	35,158.9	780.0	35,938.9	131.2	7,031.8	156.0	3,267.2
HRBs	2,304.9	69,765.7	0.0	69,765.7	384.1	13,953.1	0.0	6,342.3
CDBGs	0.0	0.0	0.0	240.0	0.0	48.0	0.0	21.8
Section 8	0.0	0.0	0.0	0.0	0.0	0.0	0.0	0.0
other	0.0	44.0	1.9	45.9	0.0	8.8	0.4	4.2
total	19,074.6	136,845.4	48,037.8	191,126.3	3,179.1	27,369.1	9,607.6	17,375.1

City total

tax increment financing	86,499.9	115,377.8	195,017.1	333,425.2	14,416.7	23,075.6	39,003.4	30,311.4
UMTA	0.0	5,112.4	34,419.1	39,531.6	0.0	1,022.5	6,883.8	3,122.2
UDAGs	30,850.9	17,353.5	5,500.0	43,209.0	5,141.8	3,470.7	1,100.0	3,928.1
IRBs	38,153.5	214,562.2	88,894.1	317,726.8	6,358.9	42,912.4	17,778.8	28,884.3
HRBs	37,342.3	147,830.0	10,373.0	159,116.0	6,223.7	29,566.0	2,074.6	14,465.1
CDBGs	98,155.2	78,533.0	63,062.7	164,565.3	16,359.2	15,706.6	12,612.5	14,960.5
Section 8	297,560.4	146,126.9	59,506.7	235,208.1	49,593.4	29,225.4	11,901.3	21,382.6
other	0.0	5,112.4	34,419.1	39,531.6	0.0	1,022.5	6,883.8	3,122.2
total	588,562.2	729,795.8	456,772.7	1,297,682.0	98,093.7	145,959.2	91,354.5	117,971.1

SOURCE: MCDA (1991a).
NOTE: See text for complete program names.

longer offer these bonds at below-market interest rates. From 1981 through 1985, developers saved an average of $72.5 million annually in interest payments by using tax-exempt revenue bonds. From 1986 through 1990, following enactment of the Tax Reform Act, average annual developer savings from these bonds dropped to less than $20 million (MCDA, 1991a).

As direct and indirect federal assistance diminished, Minneapolis has become increasingly reliant on alternative own-source revenues to fund development projects. Its primary and most controversial such revenue source is tax increment financing. In absolute terms, average annual TIF revenues have risen from $14.4 million in 1975 through 1980 to $23.1 million in 1981 through 1985, to $39.0 million in 1986-1990 (Table 6.4). In relative terms, TIF has grown from 15% of total development expenditures in 1975 through 1980 to 16% in 1981 through 1985, to 43% in 1986 through 1990 (MCDA, 1991a).

The MCDA and its predecessor agencies have earmarked different funds for specific types of projects. Revenue from federal and state governments is used almost exclusively for neighborhood projects, and tax increment funds have been targeted more to downtown and riverfront projects than to neighborhood projects. The city has allocated essentially 100% of its CDBG and Section 8 funds to residential neighborhoods. The now-defunct UDAG program was the only major source of federal funding used to support downtown development projects. The downtown area received a total of $19.2 million in UDAG money during the life of the program, 44% of the city's total UDAG revenue. The neighborhoods received essentially the same level of assistance from UDAG: $19.5 million (MCDA, 1991a).

TIF revenue, on the other hand, has not flowed to neighborhoods to the same extent as have direct federal funds. During the 1980-1990 period, neighborhood projects accounted for about one-third of total TIF revenue, with the downtown receiving almost half of the total and the riverfront one-fifth. Since 1986, downtown has received half of the city's total TIF revenue, with the neighborhoods and the riverfront each taking in one-quarter of the citywide total.

As for IRBs and HRBs, the city's other major revenue source until the Tax Reform Act of 1986, the downtown area accounted for 36% of the former and 23% of the latter in the 1980-1990 period. In the same period, neighborhoods benefited from 53% of total IRB and 33% of total HRB revenue. The riverfront received 21% of the former and 44% of the latter.

The loss of federal subsidies has been felt most acutely at the neighborhood level. Because the city has relied historically on federal housing (Section 8) and CDBG funds to support neighborhood projects, federal cutbacks have reduced the amount of money available for neighborhood development. As a result, neighborhood expenditures have plummeted since 1985. After rising by only 3% to an annual average of $85.2 million in the 1981-1985 period from $82.6 million in 1975-1980, annual neighborhood spending dropped by 38% to an average of $52.9 million in 1986-1990.

Riverfront spending declined in the post-1985 period by an even higher rate of 65%. However, at an annual average of $9.6 million, riverfront expenditures remain well above the $3.2 million annual average of 1975 through 1980. Downtown expenditures have also declined since 1986, by 14%. Compared with average annual spending in 1975-1980, expenditures in the most recent 1986-1991 period for the downtown and riverfront areas have increased markedly—twofold and threefold, respectively; only neighborhood spending has actually decreased.

TAX INCREMENT FINANCING

Minneapolis, like many localities in Minnesota and in other states, has become increasingly reliant on tax increment financing. TIF works by enabling municipalities and other local governments to use additional property taxes generated by new development to finance certain development costs. Authorized government agencies establish specific districts within which infrastructure improvements, site assembly, and other development costs are financed by increases in the district's tax base following the new development. Once created, all of the tax revenue generated by any increase in the district's tax base is collected by the authorized government authority for use within the district's project area. All other taxing authorities, such as counties and school districts, continue to collect property tax revenue generated from the district's original tax base, but not from the additional tax base created since the district was established. Once a TIF district is created, the municipality receives 100% of the tax revenues generated by increases in the tax base instead of its customary portion, which is typically less than 50% of the total levy. The justification for this arrangement is that other taxing authorities would not receive the additional revenues from the enhanced tax base had the city not initiated the development project in the first place with tax increment financing. Other governmental units ultimately benefit when the TIF district expires, usually after 15 to 30 years, when bond obligations are retired, and their tax base is augmented by the value of TIF-engendered development (Huddleston, 1984; Klemanski, 1990; Michael, 1990; Minnesota State Office of the Legislative Auditor, 1986; Minter, 1991; Paetsch & Dahlstrom, 1990). Localities usually issue bonds to provide immediate revenue for site assembly, improvement, and other costs within the area covered by the tax increment district and use tax increment revenue to cover the interest and principal payments. Because TIF works by financing new development from the tax revenue generated by the net additional value of projects financed by the TIF district, TIF is usually feasible only for high-density projects. It is more difficult to use TIF for low-density housing and commercial development because the value of the new buildings may not be sufficiently greater than that of the site's original structures to produce enough revenue to pay for the project's development costs. This problem is compounded when the TIF district must provide for "capitalized interest," that is, issue bonds large enough to pay for interest payments in the first years of the project when it has yet to yield sufficient tax capacity to pay its way (Michael, 1990). It is particularly difficult to use tax increment financing for low-density residential projects in Minneapolis because of the city's progressive property tax rates. Residential property is taxed at a much lower rate than commercial and industrial property and therefore yields less tax revenue.

Minneapolis created its first TIF district in 1971. By 1992 the city had initiated 45 such districts, of which 6 had been decertified. The tax base in all but 2 of the city's TIF districts has increased since the districts were established. TIF districts, as of 1990, accounted for 13% of the city's total tax base and generated about $62.5 million in annual revenue (Minneapolis, 1992). More than half of the city's TIF districts, and all of its most financially successful ones, are located downtown. Downtown projects were mostly established in response to proposals by private developers. Neighborhood districts, on the other hand, have usually been initiated on the suggestion of the city council. The TIF districts generating the bulk of the city's tax increment revenue were established by 1979 and have greatly benefited from the inflation of downtown property values. All but one of the city's TIF districts are controlled by the MCDA.[6]

The city's first TIF districts were established for former urban renewal areas, such as Loring Park, that had remained undeveloped by the time federal funds dried up. These projects tended to be very large and often took several years before new development began to yield tax increments. During this first period, the city collected 100% of the tax revenue generated from increases to the tax base of the entire area, whether or not the development was actually assisted by TIF funds. In 1979 the State Legislature imposed several restrictions on the use of TIF that limited the ability of TIF districts to generate revenue from tax increments that were not directly stimulated by TIF subsidies (Michael, 1990; Minnesota State Office of the Legislative Auditor, 1986).

Tax increment financing proliferated throughout the 1980s. One reason for its popularity was the growing paucity of alternative funding sources. Another is that cities considered TIF to be cost-free, because it did not involve direct city expenditures. By establishing TIF districts, cities collected the total property tax imposed on future increments to the tax base, freezing out other overlapping taxing authorities, such as counties and school districts. The latter continued to collect taxes on the district's original assessed valuation but not on any subsequent increases to the tax base.

Although TIF has become extremely popular among Minnesota municipalities, cities and suburbs both, it has been received with far less enthusiasm by counties and other taxing authorities. These governmental bodies have come to regard TIF as little more than a tax grab by municipal governments. TIF districts, they argue, deprive other government entities of revenue they might otherwise receive, thereby requiring them to levy higher tax rates on the remaining tax base than would have been necessary if their potential tax base was not diminished by TIF districts. County governments, including Hennepin County, are especially critical of TIF because they are most directly affected. The tax base for school districts is also frozen by TIF, but any resulting loss in tax revenue is largely reimbursed by state education aid. A significant portion of the cost of TIF is therefore borne by the state government when it increases its aid to local school districts to compensate for revenues they would otherwise receive (Michael, 1990; Minnesota State Office of the Legislative Auditor, 1986).

County officials do not claim to oppose tax increment financing per se. They acknowledge that the additional tax base often would not exist without public assistance provided through TIF. In some cases, however, they argue that new developments within TIF districts would have been built anyway without this assistance. When this occurs, the county is cut off from tax revenue it would have otherwise received. Although state law does stipulate a "but for" provision against funding property improvements that would be feasible without subsidy, counties believe cities and especially suburbs have not always followed the spirit if not the letter of these regulations. Counties are especially vexed by the use of TIF districts to subsidize the construction of suburban shopping centers and to fund roads and other infrastructure improvements that would normally be funded through special assessments or gasoline taxes (Hennepin County Board of Commissioners, 1990; Hennepin County Staff, 1984; Minnesota State Office of the Legislative Auditor, 1986).

Counties object perhaps even more vociferously to the time period in which TIF districts remain in effect. They argue that TIF bonds could be retired sooner, and that TIF districts should not be so easily able to fund additional activities within their project areas, thus prolonging the life of the district and postponing when tax increments are returned to the county tax base (Hennepin County Board of Commissioners, 1990; Hennepin County Staff, 1984).

In response to these criticisms, the Minnesota State Legislature restricted the use of TIF in 1990. Most important, the Legislature reduced state aid to local governments to reflect the captured tax increment revenue that is otherwise excluded by state aid formula calculations. By lowering state aid in proportion to a locality's reliance on TIF, the state effectively put a portion of TIF's cost on the city budget. To make up for state aid lost because of TIF, municipalities are forced to either increase taxes or cut their expenses; TIF is no longer cost-free. This provision applies only to tax increment districts established after 1990; it does not affect existing districts. Other restrictions imposed in 1990 limit the amount of time parcels can remain within a district before undergoing any improvements, the extent to which revenues can be pooled from two or more districts, and the uses for which tax increment revenue may be spent (Michael, 1990).

According to the MCDA, current state regulations make it impossible for the city to establish new TIF districts. Creating new districts, as noted above, would cause state aid to be reduced, forcing the city to either raise taxes or cut expenditures, neither of which is realistic. Tax rates in Minneapolis and throughout Minnesota are considered extremely high, and it is politically unacceptable to raise them further. The MCDA has been lobbying the state legislature to repeal TIF regulations affecting state aid and to loosen several TIF restrictions.

Neighborhood Revitalization Program

Minneapolis's development priorities began to shift from the downtown to residential neighborhoods in the late 1980s. Criticism mounted that the city government, especially the MCDA, was spending an undue proportion of its resources on downtown development, to the detriment of the neighborhoods. A survey conducted by the City Planning Department, for example, revealed widespread discontent among city residents toward downtown development. By decade's end, several neighborhood advocates had been elected to the city council. These new representatives helped convince the council, which is also the board of directors of the MCDA, to redirect the city's development strategy.

The cornerstone of the city's new development strategy is the Neighborhood Revitalization Program (NRP). Under this 20-year, $400 million program, each of the city's 81 neighborhoods will have the opportunity to develop, in public meetings, its own revitalization plan and specify its priorities for housing and public services. These neighborhood plans must be approved by various governmental authorities. The program provides $20 million annually to assist neighborhood projects, about half of which must be spent on housing. In addition, the program requires city and county agencies as well as the school district to incorporate neighborhood plans into their capital budgets. The program was approved by the Minneapolis City Council and by the Minnesota State Legislature in June 1990 (Fainstein, 1992). As of October 1993, 32 of the city's 81 neighborhoods had initiated a series of workshops to formulate neighborhood revitalization plans. Six neighborhoods had completed their plans, which were in various stages of implementation (Fainstein, 1993).

The NRP is funded by the city's tax increment districts. In essence, revenue from downtown development projects assisted with tax increment financing is channeled to the neighborhoods via the NRP. The NRP is funded through the Common Development and Redevelopment Project (known as the Common Project), created in December 1989 "to link the financial success of the Downtown tax increment districts with the critical redevelopment needs of Minneapolis neighborhoods" (MCDA, 1992, p. 1). The Common Project was established by combining most of the city's existing redevelopment project areas and development districts. Most of the city's existing tax increment project areas were consolidated into a single project area to allow revenue generated from successful downtown projects to be spent in needier areas. The Common

Project is financed in part by refunding most of the city's tax increment bonds at a lower interest rate over a longer time period, thereby reducing annual interest costs. Additional funding for the Common Project is derived from other tax increment revenue and from the MCDA's development accounts, which consist of revenue generated from land sales, lease income, loan repayments, and parking fees (see MCDA, 1988). In addition to the $20 million cost of the NRP program, the Common Project also funds up to $20 million a year for other neighborhood development activities, and contributes approximately $7.5 million each to the county and the school district to help them meet their obligations to the NRP program. These contributions serve to compensate these government authorities for the loss of tax revenue they would have received had the terms of existing tax increment bonds not been extended.

The city's use of downtown development projects to fund neighborhood revitalization is somewhat similar to the more common use of inclusionary zoning and linkage programs. These programs require private developers to provide low-cost housing, either directly or by contributing to a housing trust fund, in exchange for the opportunity to construct new residential or nonresidential development, with the amount of the exaction proportional to the size of the development (Mallach, 1984; Nenno & Colyer, 1990). The Minneapolis NRP program differs from inclusionary zoning and linkage programs in that it is based on the accumulated value of private development (the tax base within tax increment districts), not on year-to-year investments. It is therefore less vulnerable to the vicissitudes of the real estate market; revenue does not rise and fall in lockstep with the amount of new development constructed each year.

Development Impacts

No one in government, academia, or elsewhere is known to have conducted any assessments of the economic, social, fiscal, or other impacts of Minneapolis's redevelopment. To paraphrase a local observer, impact studies are done when projects are proposed, not after they are completed. Nevertheless, redevelopment has had obvious effects in some areas, such as in the tax base and in the city's nightlife, whereas impacts are less certain in other areas, especially retail.

TAX BASE

The effect of downtown redevelopment is probably most evident in changes in the tax base of the city and of the entire metropolitan region. Since 1980, downtown development has totaled nearly $4 billion in new construction. As a result of this development, the downtown area's share of the city's tax base almost doubled, from 21% in 1980 to 41% in 1991 (MCDA, 1991b).

Even more dramatic is the effect on the region's commercial-industrial tax base and on the distribution of this tax base under the Fiscal Disparities program. As described earlier, the Fiscal Disparities program channels 40% of each municipality's post-1971 C-I tax base into a regional pool, which is then redistributed to individual municipalities according to their populations and tax capacities. When the program started in 1975, Minneapolis was the second-largest net recipient, receiving $10.8 million more than the $16.9 million it contributed to the regional pool. The only city to receive more on a net basis was St. Paul. In 1980, Minneapolis was the largest net recipient, St. Paul second; by 1985, however, downtown development turned Minneapolis into the largest net contributor, and it received $22.4 million less than the $238 million it contributed to the program. The city continued to be the program's largest net contributor through 1991. St. Paul, on the other hand, remains the program's largest net recipient. No other city in the metropolitan region program has transformed itself from a major net recipient to a major net contributor (Metropolitan Council, 1991).

Like many cities, Minneapolis has faced declines in property values since the late 1980s. Although vacancy rates for commercial property have declined from more than 20% in 1992 to approximately 12% in 1994, commercial property values are 25% below their peak in 1989. The diminishing tax base has hurt TIF revenue, which has decreased by about $10 million (16%) from 1993 to 1994 (Mark Garner, planner, MCDA, interview, March 1994). With development expenditures increasingly tied to tax increment financing, decreases in the tax base erode the amount of "captured" tax revenue available for debt payments and other development costs.

NIGHTLIFE AND OTHER ENTERTAINMENT

Partly as a result of its redevelopment, downtown Minneapolis no longer seems to close down at the end of the workday. With the downtown housing stock increased, several new hotels, several new shopping centers, two new sports facilities, a major convention center, and restoration of major theaters, not to mention scores of new restaurants, downtown Minneapolis is busy well into the night. Together with the nearby warehouse district and the riverfront area, downtown Minneapolis offers residents of the metropolitan area and tourists an expanding array of leisure activities—sports events, concerts, galleries, restaurants, and theater.

The improvement of downtown nightlife was noted immediately by most of the academic, government, and business observers interviewed for this study, but it is difficult to quantify. A partial measure of this development is afforded by employment data in the 1980-1988 period for eating and drinking places and for amusement and recreation services. Unlike most other sectors, the city's employment growth in these two industries kept pace with the rest of the metropolitan region (Minneapolis, 1990).

RETAIL

The preservation and enhancement of the downtown shopping district has long been a leading objective of the city's redevelopment activity. This was the prime motivation for remaking the Nicollet Mall in the 1960s and for subsidizing the construction of several upscale shopping centers in the late 1980s. The impacts of these efforts on the regional retail market, however, are inconclusive at best. The Census of Retail Trade shows that the city's share of the metropolitan area's total retail sales declined from 46% in 1958 to 31% in 1967, to 17% in 1977, to 14% in 1987, the latest year available. The central business district's share of total regional shares dropped from 16% in 1958 to 10% in 1967, to 5% in 1977, and to 4% in 1982, the last year in which the Census Bureau compiled CBD sales data (Ashwood, 1991, p. 299). Ironically, the downtown's sharpest decline in retail sales and market share occurred from 1967 through 1977, the decade immediately following the opening of the Nicollet Mall.

The retail situation in Minneapolis seems to have stabilized since 1977. The city's share of total regional sales dropped by just 3 percentage points in the 1977-1987 period (and the CBD's share by only 1 point in the 1977-1982 period). With the subsequent opening of two major downtown malls anchored by Saks Fifth Avenue and Neiman Marcus, the 1992 Census of Retail Trade could show some improvement in the city's retail market.

The arrival of Saks Fifth Avenue and Neiman Marcus in downtown Minneapolis has not increased the number of downtown department stores. Rather, these upscale retailers replaced two more moderately priced department stores (Powers and J. C. Penney) that had closed in the mid-1980s. A third store, Carson Pirie Scott, closed in January 1993. Downtown is increasingly considered an exclusive district of high-end retailers, offering little for low- and moderate-income customers. The city government (MCDA) and the Downtown Council believed it was essential to attract luxury retailers, such as Neiman Marcus, downtown in order to distinguish it from suburban shopping centers. These retailers do not operate any other stores in the metropolitan region outside downtown Minneapolis. The city and the Downtown Council recognize that the downtown is lacking in stores with mass-market appeal. To help fill this gap, they recruited several moderate-priced retailers, including Filene's Basement and Montgomery Ward, to occupy portions of the space vacated by Carson Pirie Scott.

Whatever gains downtown retailers have made in the last few years, however, could well be wiped out by intensified suburban competition. The nation's largest shopping center, the Mall of the Americas, opened on August 11, 1992, in suburban Bloomington, Minnesota, approximately 10 miles south of downtown Minneapolis. Known locally as the megamall, the $805 million, 4.2-million-square-foot Mall of the Americas contains only 500,000 square feet less retail space than all of downtown Minneapolis. It features four anchor department stores, 1.7 million square feet of additional retail space, an indoor amusement park, a 14-screen movie theater, several nightclubs and restaurants, an 18-hole miniature golf course, and a 300-foot walk-through aquarium ("Guys and Malls," 1992; MarksJarvis, Kahn, & Parker, 1992).

One year after its opening, the mall had met its projection of more than 35 million visits and more than $600 million in sales (Rhees, 1993; Wieffering, 1994, p. 13). Overall, the mall's impact on regional retailing has been less severe than initially feared. Partly because 30% to 40% of its customers are tourists and because it introduced three new department stores as well as an amusement park to the metropolitan region, the megamall has expanded the region's retail market more than it has realigned the previously existing market. Retailers at Minneapolis shopping centers claim that sales have rebounded after declining immediately after the mall first opened. Downtown St. Paul, on the other hand, has suffered from the mall's competition (Wieffering, 1994, p. 14).

DISPLACEMENT

Urban redevelopment almost always displaces the occupants—households or businesses—of the buildings demolished or rehabilitated to make way for the new. Only when already vacant structures are redeveloped does direct displacement not occur. As noted earlier, the city's Gateway redevelopment project of the 1950s and 1960s displaced more than 2,000 people and several hundred businesses. Subsequent downtown redevelopment projects, especially Loring Park and the Convention Center, have also uprooted many downtown residents and establishments. The extent of residential displacement has been limited, however, because relatively few people have resided in the downtown area.

Policy Implications and Conclusions

Minneapolis's surge of development since 1980 is essentially over. A few projects still await action, particularly on Hennepin Avenue, but there is scant political or financial support for more downtown development. With the present surplus of office and retail space, city leaders agree it would be foolish to continue building as before. Development priorities have shifted. For downtown, the focus has turned to marketing—attracting visitors and especially businesses to the remade downtown.

The redevelopment of downtown Minneapolis exemplifies the capacity of local government to stimulate commercial and residential development. With dwindling federal dollars and the creative use of alternative revenue sources, especially tax increment financing, the city has managed to revamp the downtown area by rebuilding its skyline and diversifying its economy and activities. Even more impressive, the city has managed to tap the accumulated growth in the downtown tax base to fund a long-term neighborhood revitalization program.

Although the redevelopment of Minneapolis is exemplary in several respects, it may prove to be a difficult model for other cities to adopt. The city's relative affluence, its large number of major corporations, its racial homogeneity, and the apparent public trust in government most likely made it easier for the city administration to pursue its redevelopment agenda than would be the case in other large cities. Equally important, the city has, until recently, almost always succeeded in getting the Minnesota State Legislature to approve critical measures, such as the use of tax increment financing to subsidize the citywide Neighborhood Revitalization Program. Few other cities are so fortunate in their dealings with state government.

Minneapolis also illustrates how cities are constrained in their ability to foster downtown development and neighborhood revitalization. Long-standing conflicts with the county and, to a lesser extent, other government authorities over the use of tax increment financing has led to increased restrictions over the future application of this key revenue source. At present, the city considers it virtually impossible to establish new TIF districts because of the resulting loss of state aid. In addition to intergovernmental constraints, the city faces even more fundamental economic limits. The longest postwar recession, coupled with a serious downturn in real estate values, has weakened the city's tax base—the basis for tax increment financing. Falling commercial property values have reduced the MCDA's cash-flow projections below its most conservative previous estimates. So far, the $20 million annual budget for the NRP program appears secure, at least for the next few years, but there is little margin for error. Further erosion of the tax base could undermine the city's effort to marshal its own resources to fund comprehensive neighborhood revitalization.

If Minneapolis, with its prosperous economy, homogeneous population, and relatively harmonious relationship with the state government, cannot rely on its own resources to sustain downtown and neighborhood development, it is not reasonable to expect the same of any other major city. Without federal and state assistance, Minneapolis and other cities are unable to foster downtown and especially community development. This is particularly true of low- and moderate-income neighborhoods, which are by themselves unable to generate tax revenues necessary to support development projects. Community Development Block Grants and project-based housing subsidies such as the Section 8 New Construction program are critical to the funding of neighborhood development. Federal tax expenditures, such as tax-exempt industrial and housing bonds—currently prohibited by the Tax Reform Act of 1986—can also play a constructive role in local redevelopment programs, provided they do not subsidize projects that would otherwise attract private financing.

Notes

1. Income figures are presented in constant 1982-1984 dollars, based on the national Consumer Price Index for All Urban Consumers. In current dollars, per capita income in Minneapolis has increased from $3,483 in 1969 to $7,940 in 1979, to $13,092 in 1987, and median family income from $9,958 in 1970 to $19,737 in 1980, to $32,998 in 1990.

2. Industry-specific employment data for the city of Minneapolis are not available for the years since 1988.

3. For details on the allocation of Fiscal Disparities funds to individual municipalities, see Baker et al. (1991) and Metropolitan Council (1991).

4. Private donations paid for the construction of the $7 million Orchestra Hall (Ashwood, 1991).

5. One of the new hotels, the Marriott, was built as part of the City Center project. It replaced another Marriott that was located nearby.

6. The only TIF district not controlled by the MCDA was established by the city council in the late 1980s to finance the settlement of a lawsuit filed against the city by a private developer.

References

Abler R., Adams, J. S., & Borchert, J. R. (1976). *The Twin Cities of St. Paul and Minneapolis.* Cambridge, MA: Ballinger.

Anderson, B. (1990). *Growing percent giving programs: The Minnesota model.* Minneapolis: Greater Minneapolis Chamber of Commerce.

Aschman, F. T. (1971). Nicollet Mall: Civic cooperation to preserve downtown's vitality *Planner's Notebook 1*(6), 1-8.

Ashwood, D. (1991). *Downtown redevelopment: The public-private partnership, and the downtown mall solution. A comparison of the redevelopment process in Minneapolis and Milwaukee.* Unpublished doctoral dissertation, University of Wisconsin—Milwaukee.

Baker, K., Hinze, S., & Manzi, N. (1991). *Minnesota's Fiscal Disparities program.* St Paul: Minnesota House of Representatives, Research Department.

Brandl, J., & Brooks, R. (1982). Public-private cooperation for urban revitalization: The Minneapolis and Saint Paul experience. In R. S. Scott & R. A. Berger (Eds.), *Public private partnerships in American cities* (pp. 163-200). Lexington, MA: Lexington.

Breckenfeld, G. (1976, January). How Minneapolis fends off the urban crisis. *Fortune* pp. 130-135.

Fainstein, S. S. (1992). *A preliminary evaluation of the Minneapolis Neighborhood Evaluation Program.* New Brunswick, NJ: Rutgers University, Center for Urban Policy Research.

Fainstein, S. S. (1993). *An interim evaluation of the Minneapolis Neighborhood Revitalization Program.* New Brunswick, NJ: Rutgers University, Center for Urban Policy Research.

Fainstein, S. S., & Fainstein, N. I. (1991). *Public-private partnerships for urban (re)development* (Working Paper No. 35A). New Brunswick, NJ: Rutgers University Center for Urban Policy Research.

Focus [Newsletter of the Greater Minneapolis Chamber of Commerce]. (1992, July). Vol. 2, p. 2.

Fraser, D. M., & Hively, J. M. (1987). Minneapolis: The city that works. In G. Gappert (Ed.), *The future of winter cities* (pp. 131-138). Newbury Park, CA: Sage.

Galaskiewicz, J. (1985). *Social organization of an urban grants economy: A study of business philanthropy and nonprofit organizations.* New York: Academic Press.

Garner, M. (1991). *Housing the downtown worker: Housing development, new neighborhood creation and the downtown growth coalition in Minneapolis.* Unpublished manuscript, University of Minnesota, Department of Geography.

Goetz, E. G. (1992). Tax base sharing in the Twin Cities. *Urban News, 6*(1), 7-10.

Goldfield, D. (1976). Historic planning and redevelopment in Minneapolis. *Journal of the American Institute of Planners, 42,* 76-86.

Guys and malls: The Simons' crapshoot. (1992, August 17). *Business Week,* pp. 52-53.

Hauser, D. (1990, September 17). Growth has evened out the skyline over the last 20 years. *Skyway News,* p. 9.

Hennepin County Board of Commissioners. (1990). *Tax increment financing.* Minneapolis: Hennepin County.

Hennepin County Staff. (1984). *Tax increment districts: A report to the Hennepin County Board of Commissioners.* Minneapolis: Hennepin County.

Huddleston, J. R. (1984). Tax increment financing as a state development policy. *Growth and Change, 15*(2), 11-17.

Klemanski, J. S. (1990). Using tax increment financing for urban redevelopment projects. *Economic Development Quarterly, 4,* 23-28.

Klobuchar, J. (1992, January 21). Now, in its glory, Metrodome doesn't look so evil. *Minneapolis Star Tribune.*

Lassar, T. J. (1989). The pros and cons of downtown skywalks. *Urban Land, 47*(12), 2-6.

Lassar, T. J. (1991). Sharing the benefits and costs of growth management in Minneapolis. *Urban Land, 50*(2), 20-25.

Leitner, H., & Garner, M. (1993). The limits of local initiatives: A reassessment of urban entrepreneurialism for urban development. *Urban Geography, 14,* 57-77.

Lu, W. (1984). Urban design in the Twin Cities: Minneapolis. *Planning, 50*(3), 12-22.

Maki, W. (1990). *Outlook for the Minneapolis economy in the 1990s.* Report prepared for the Minneapolis Community Development Agency.

Mallach, A. (1984). *Inclusionary housing programs.* New Brunswick, NJ: Rutgers University, Center for Urban Policy Research.

MarksJarvis, G., Kahn, A., & Parker, W. (1992, April 19). Mcgamall still looking for tenants. *Saint Paul Pioneer Press,* pp. A1, A8.

Martin, J. A., & Goddard, A. (1989). *Past choices/present landscapes: The impact of urban renewal on the Twin Cities.* Minneapolis: University of Minnesota, Center for Urban and Regional Affairs.

Metropolitan Council. (1991). *Fiscal Disparities discussion paper* (staff report draft, Publication No. 620-91-066). St. Paul: Author.

Michael, J. (1990). *Tax increment financing project: A background paper.* St. Paul: Minnesota House of Representatives, Research Department.

Millet, L. (1989). Renaissance on the Minneapolis riverfront. *Inland Architect, 33*(3), 13-14.

Minneapolis, City of. (1990). *Non-agricultural employment change, 1980-1988: Minneapolis and the Minneapolis-St. Paul metropolitan area.* Minneapolis: Office of the City Coordinator.

Minneapolis, City of. (1992). *State of the city 1992.* Minneapolis: Office of the City Coordinator/Minneapolis Planning Department.

Minneapolis Community Development Agency (MCDA). (1984). *Minneapolis downtown development 1978-1983: A summary report.* Minneapolis: Author.

Minneapolis Community Development Agency (MCDA). (1986). *Minneapolis downtown development: 1985-1986 summary report.* Minneapolis: Author.

Minneapolis Community Development Agency (MCDA). (1987). *Minneapolis downtown development: 1986 summary report.* Minneapolis: Author.

Minneapolis Community Development Agency (MCDA). (1988). *Minneapolis Common Project proposal: Common development and redevelopment plan and common tax increment plan for Minneapolis Community Development Agency.* Minneapolis: Author.

Minneapolis Community Development Agency (MCDA). (1990a). *Meet the MCDA.* Minneapolis: Author.

Minneapolis Community Development Agency (MCDA). (1990b). *1989 annual report.* Minneapolis: Author.

Minneapolis Community Development Agency (MCDA). (1991a). *MCDA project and program expenditures: 1975-1990.* Minneapolis: Author.

Minneapolis Community Development Agency (MCDA). (1991b). *Minneapolis down town development: 1989-1990 summary report.* Minneapolis: Author.

Minneapolis Community Development Agency (MCDA). (1992). *City of Minneapolis tax increment resource management.* Minneapolis: Author.

Minneapolis Planning Department and Downtown Council of Minneapolis. (1988). *Metro 2000 plan: Minneapolis metro center.* Minneapolis: Authors.

Minnesota State Office of the Legislative Auditor, Program Evaluation Division. (1986). *Tax increment financing.* St. Paul: Author.

Minter, N. L. (1991). Tax increment financing. *Urban Land, 50*(4), 38-39.

Mitchell, S. (1974, September). Minneapolis really did it. *Stores,* pp. 10-12.

Nenno, M. K., & Colyer, G. (1988). *New money and new methods: A catalog of state and local initiatives in housing and community development.* Washington, DC: National Association of Housing and Redevelopment Officials.

Ouchi, W. G. (1984). *The M-form society.* Reading, MA: Addison-Wesley.

Paetsch, J. R., & Dahlstrom, R. K. (1990). Tax increment financing: What it is and how it works. In R. D. Bingham, E. W. Hill, & S. B. White (Eds.), *Financing economic development* (pp. 83-97). Newbury Park, CA: Sage.

Rhees, S. S. (1993, October). Mall wonder. *Planning, 59,* 18-23.

Roe, L. K., & Rucker, R. E. (1991). Human ecology and urban revitalization: A case study of the Minneapolis warehouse district. *Mid-American Review of Sociology, 15*(1) 1-16.

Wallace, D. J. (1992, November 11). New hotel in Minneapolis. *New York Times,* p. D17.

Wiedenhoeft, R. (1975). Minneapolis: A closer look. *Urban Land, 34*(9), 8-17.

Wieffering, E. (1994, February). What has the Mall of the Americas done to Minneapolis *American Demographics,* pp. 13-16.

Worthy, F. S. (1987, January 19). The could-do city could do it again. *Fortune,* pp. 82-90.

Conclusion

It is evident that of perhaps equal importance to the identification of successful revitalization strategies is the examination of the preceding case studies for the factors that promoted the strategies' success. There are, after all, many cities across the nation that have tried some of the strategies identified by the contributors to this volume, many of which have not met with success. Indeed, the case studies reveal that some of the strategies used in the seven subject cities did not lead to success. What conditions were present that contributed to success or were necessary precursors to success?

Keys to Successful Revitalization

Strong Public Leadership. Leadership was an element in virtually every case study of a successful revitalization program. This leadership came from elected officials, particularly mayors, as well as from professionals on planning and development staffs. Strong leadership involves a variety of factors: clear vision of the future of the city, region, or project; strategic placement to expedite and move projects; strong motivation to overcome the inevitable hurdles that arise in all complex projects; knowledge about the tremendous variety of development tools and programs available; and ability to select among those tools the ones most appropriate for the community. For example, Brooklyn Borough President Howard Golden and his staff took a more aggressive role in redevelopment efforts than had previous presidents and their administrations, eventually successfully mediating between community groups and developers to prevent blockage of controversial projects.

Under the category of leadership could also be listed the leadership of community-based organizations and even private development leadership. Many times, strong leadership in these organizations is also critical to success.

Well-Focused Planning Concepts. Many of the case studies presented here reveal the existence of well-thought-out, well-focused plans at the base of successful redevelopment projects or strategies. Although it is clear that not all aspects of every plan will be carried out over a long period of time, the existence of a strong plan will identify and direct strategies in a way that maximizes their effect. In New Orleans for example, the existence of a general consensus plan for the central business district since 1975 has led to the implementation of a wide variety of projects that support one another and that have created a very successful downtown.

The Ability to Respond to Traumatic Events. As we mentioned in our introduction to this volume, the ability to respond to traumatic event has often been a necessary ingredient in galvanizing action to address a particular problem. It usually takes strong, effective leadership to respond at times of crises and to take advantage of such situations.

Existing Community Characteristics. A community's characteristic have much to do with the effectiveness of any strategy. Communities that are relatively affluent and well educated and have strong business leadership seem more apt to implement strategies successfully, particularly unique and unusual strategies. These conditions are identified as important elements in the ability of Portland and Minneapolis to implement regional growth management and regional tax base sharing, both strategies in limited use around the country. In those communities the do not have these basic characteristics, leaders may have to work harder to educate their populace about programs they wish to implement and to develop the cooperation and knowledge necessary to see them through to completion.

Good Relationships Between Levels of Government. Good relationships, whether city-state relationships or city-county relationships, are a prerequisite of success. For example, the Minneapolis Downtown Council had to work with local, state, and federal government agencies to get financing and regulatory approval for its Nicollet Mall project. In the absence of such relationships, hostile action by a county council or state legislature can make the use of many local strategies difficult.

The factors noted above appear, in whole or in part, to contribute greatly to the probability of success of revitalization efforts; indeed, some may be absolute necessities. Officials planning revitalization efforts should be knowledgeable about these factors and should work to have them present as a basis for the success of their programs.

Key Impediments to Successful Revitalization

Just as the case studies in this volume identify factors critical to the successful implementation of revitalization strategies, they indicate factors that inhibit revitalization success as well:

Condition of the Local and National Economy. Economic conditions are a major factor affecting virtually any type of strategy. Indeed, where the economy is weak or even in a recession, successful implementation of programs will be very difficult. The ability to identify the exact condition of the economy appears critical to success. Obviously, the strategies needed for revitalization efforts in a depressed or declining economy may be different from those needed in one that is relatively healthy.

Cuts in Federal Aid. Federal cutbacks have been devastating to the ability of cities to implement investment strategies. This is particularly important in relation to individual project financing, where federal aid such as economic development grants and Urban Development Action Grant programs have played such a large role. The well-documented retreat of the federal government from an active role in the economic development of central cities during the 1980s has greatly inhibited successful revitalization projects.

Recent Legislation. Recent federal legislation, such as the Tax Reform Act of 1986, and specific state legislation limiting the ability of several cities, including Minneapolis, to use tax increment financing has inhibited a variety of revitalization efforts. These actions by legislative bodies have the effect of stifling the economic and legal climate necessary for revitalization.

Policy Implications and Recommendations

What, then, are the overall implications of the case studies presented here, and what can federal, state, and local policymakers learn from the lessons taught by these cities' experiences? The studies' overall findings may be grouped into six distinct areas:

1. leadership capacity building
2. promotion of regional cooperation and problem solving
3. federal financial commitment to central-city revitalization programs
4. human investment strategies
5. lessons for federal interagency cooperation
6. federal support for local planning efforts

Discussion of each of these areas and their impacts on future policy directions follows.

LEADERSHIP CAPACITY BUILDING

It is clear from an analysis of the case studies that informed, visionary leadership is key to the effectiveness of any revitalization strategy. This includes elected as well as professional leadership and may include community leadership as well. How does one ensure that such informed leadership is available when it becomes time to steward polices and programs? It seems apparent that programs that build the capacity for local leadership will make a fundamental difference in revitalization programs

This capacity building at the local level should take on a variety of forms. Training programs to inform elected officials, particularly chief executives and city council members, of the latest revitalization techniques in use throughout the country would be very useful. Many revitalization strategies are probably not utilized because key decision makers simply do not know about them. Training in how these programs work and access to examples of their success or failure in real city situations would give elected officials a greater grasp of the tools available to them. Although a variety of training programs exist, a complete inventory of these elected official training programs should be undertaken by the federal government, a comprehensive elected leadership curriculum created, and institutions funded to carry out the program. This task is ideal for federal intervention, because it is in the direct interest of the federal government to ensure that local elected officials receive the most comprehensive training. This will certainly build many more informed leaders, which will directly improve the chances of success for revitalization programs.

Similarly, training for professional staff delegated to manage and carry out these strategies is also imperative. A complete analysis of the content of existing programs, their constituencies, and their effectiveness would undoubtedly lead to better-designed programs for senior professional staff. A program for local officials similar to that the federal government provides for senior staff of federal agencies would greatly increase local capacity to implement complex redevelopment strategies.

Training for key personnel within community-based organizations would greatly increase their capacity to help government officials utilize complex strategies by building support within local communities and giving community leaders the information necessary to modify programs to suit particular local needs.

The federal government could take a leadership role in surveying current training efforts in all three of the areas described above, in bringing order to these programs, and in developing new programs to fill obvious training gaps. This effort at building local leadership capacity to implement successful revitalization strategies could be among the most cost-effective expenditures concerning development and revitalization made by the federal government.

PROMOTION OF REGIONAL
COOPERATION AND PROBLEM SOLVING

Growing evidence in the NCRCC studies and others points to the fact that regional planning approaches to problem solving have greatly aided central-city revitalization. In fact, the evidence of the case study comparing Atlanta and Portland suggests that unmanaged regional development will put revitalization of central cities at a distinct disadvantage. This is a sobering thought for policymakers at all levels to contemplate. Are federal investments in revitalization of central cities seriously threatened in those cities that cannot or will not manage the growth of the entire urban area? Clearly, the success of such cities as Portland and Minneapolis, which do exercise strong regional planning, regional service delivery, and regional growth control, indicate that there are important lessons to be learned.

Recent passage of the new federal intermodal transportation bill, which greatly expands the role of regional planning in transportation decisions, suggests that the federal government has begun to reassert its demands for a regional perspective to problems and fund allocations that are by their very nature regional in scope. The federal government should aggressively pursue these questions and try to determine whether a direct correlation exists between effective regionalism and effective central-city revitalization efforts. In the meantime, evidence suggests that regional planning efforts should be aggressively promoted and cooperative regional efforts required wherever feasible. The case studies further suggest that efforts to support regional planning must be meaningful and firm in order to be effective. Such programs advanced by the federal government must have teeth and must strongly encourage—and in some cases force—regional cooperation. There will remain few examples of true regional cooperation without such forcefulness.

FEDERAL FINANCIAL COMMITMENTS
TO CENTRAL-CITY REVITALIZATION PROGRAMS

Analyses of central-city revitalization programs are replete with examples of successful projects made possible, all or in part, by federal financial assistance. They are also replete with examples of programs that failed or were scaled back when federal assistance was withdrawn. It is clear that many important revitalization efforts cannot be undertaken by local communities without substantial federal assistance. Much of the decade of the 1980s saw a retreat in this role of the federal government and NCRCC research makes it evident that such a retreat has immeasurably hurt central cities. The federal government cannot hope to see major improvements in central cities if it withdraws from the roles of facilitator and builder of public infrastructure. In one way or another, the case studies from all seven subject cities confirm this idea.

No federal purpose can be served by the decline of central cities. Indeed, it might be argued that the decline in central cities represents the greatest danger to our national defense as we weaken from within. A wholesale reevaluation of the investment strategy used by the national government should be undertaken, with an eye to those programs that most effectively spur job creation and development.

HUMAN INVESTMENT PROGRAMS

It is clear that physical redevelopment alone will not bring prosperity to central cities. Much greater attention must be paid to investing in human capital as an effective strategy for long-term economic improvement. The federal government must ensure that its programs, in whole or in part, encourage training and educational efforts designed to increase the skills of workers so they are ready for the economic opportunities of the twenty-first century. Federal regulations that currently limit the effectiveness of existing programs must be changed. The damage done by such regulations is evident in the Baltimore JOBS program. The initial version, which was touted as being very successful, was replaced by a version that required more bodies to pass through the system in order to meet federal participation rate requirements and thus receive funding.

New human resource strategies must be constantly developed, and new forms of partnerships among business, government, and educational institutions must be developed as well. Human resource strategies should be a part of almost all federal economic development programs and should be a part of all local planning efforts. Just as building capacity for local leadership may be the most cost-effective strategy for ensuring successful revitalization programs at the local level, investment in people may yet prove the most important single investment federal, state, or local governments can make.

FEDERAL INTERAGENCY COOPERATION

The Baltimore and Fort Worth case studies suggest that there are at least two major problems with cooperation among federal agencies that have negative impacts on central cities. First, benefit programs are written separately, without regard to how different programs affect the recipients. Rules that limit benefits on one side of the ledger also hurt eligibility requirements on the other. The federal government must ensure that its many components communicate with one another in order to ensure a unified and favorable impact on the target populations. Second, the federal government does not seem to be able to look at a single central city from a coordinated viewpoint. There is no coordinated strategy for directing aid or particular types of aid to a particular city. The inability of the federal government to coordinate its programs on a city-by-city basis prevents the maximum benefit of each program from being achieved.

SUPPORT FOR LOCAL PLANNING

All of the case studies presented in this volume are remarkably clear on one final point: Successful redevelopment strategies and programs have all developed from well-thought-out plans and economic development strategies. It is clear that those communities that invest in the process of planning and produce cogent, comprehensive plans for their communities have greater chances for successful revitalization than do those communities that do not. Therefore, the federal government should support comprehensive and strategic planning at the local level. It should do this both through mandated planning requirements in its programs (the new transportation bill, for example) and in its financial support for planning efforts. Once again, it can be argued that the effectiveness of federal resources is jeopardized when adequate planning is not accomplished. Federal funding to communities to support planning should be examined in order to give revitalization programs the best chance for success.

The implications and conclusions of the case studies are clear and give the promise of more effective local efforts. It is our hope that policymakers at all levels will digest these lessons and put into policy both the ingredients that foster success and the particular revitalization programs that have proven themselves successful.

Future Directions

The NCRCC case studies presented in this book go as far as available resources allowed. In the fall of 1994, the NCRCC was refunded by

Congress for $1 million; it continues to pursue its research, clearinghouse, and applied technical assistance activities on behalf of the Congress, the administration, and individual cities. The U.S. Department of Housing and Urban Development Office of Research and Policy Development and the NCRCC are jointly pursuing an ambitious urban revitalization research agenda. NCRCC researchers have identified many related areas of inquiry that should be pursued, including the following:

1. investigation of other revitalization programs in the current case study cities
2. investigation of international cities—their continued vitality and their revitalization programs
3. determination of what effects current revitalization programs have on the federal deficit
4. continued analysis of regional urban growth management programs and their effects on central-city revitalization efforts
5. analysis of subsidies granted to suburban communities and the effects they have on support for central-city and regional urban form
6. investigation of the relationships among urban form, central-city vitality, regional economic vitality, and national competitiveness
7. investigation into regional governmental mechanisms
8. research into a variety of topics on urban form
9. documentation of the requirements for ongoing monitoring of revitalization projects
10. further explanation of the role of urban design in central-city revitalization efforts
11. investigation of federal tax and loan policies that work against neighborhood revitalization projects
12. examination of the unintended consequences of federal legislation for revitalization efforts
13. research on service area boundaries and various bonding programs
14. investigation into the effects of revitalization efforts on a variety of target groups

The revitalization of America's central cities is an extremely complex, challenging problem. At this time in the nation's history, there is perhaps no weightier domestic item on our agenda.

FRITZ W. WAGNER
TIMOTHY E. JODER
ANTHONY J. MUMPHREY JR.

Name Index

212

Subject Index

216

About the Editors

Fritz W. Wagner is Dean of the College of Urban and Public Affairs at the University of New Orleans, a post he has held since 1988. He is also Director of the National Center for the Revitalization of Central Cities. He received his undergraduate degree from Michigan State University and his master's degree (1971) and Ph.D. (1974) from the University of Washington. He joined the faculty of the University of New Orleans after completing his graduate work, as Assistant Professor in the Urban Studies Institute (now the College of Urban and Public Affairs), and has spent his entire university career, now 20 years, at that university. His research interests are in the areas of revitalizing central cities and small towns and urban recreation planning.

Timothy E. Joder is Director of the Louisiana Urban Technical Assistance Center, the professional public service arm of the College of Urban and Public Affairs at the University of New Orleans. Concurrently, he serves as Deputy Director of the National Center for the Revitalization of Central Cities, a consortium of university researchers across the nation who are attempting to provide the federal government with prescriptive advice on national urban policy. Prior to joining the University of New Orleans in 1981, he served as Chief Planner, State Planning Office, Office of the Governor of Louisiana; as a community development planner with the city of Baton Rouge and East Baton Rouge Parish; and as coordinator of the Boca Raton, Florida, Community Development Block Grant program. He holds a B.A. in urban affairs/geography from the University of Pittsburgh and the M.P.A. degree from the University of New Orleans.

Anthony J. Mumphrey Jr. is President of the Mumphrey Group, Inc., and Professor of Urban and Regional Planning at the College of Urban and Public Affairs of the University of New Orleans. He began his professional planning career in 1968 at the University of New Orleans, where he served as Research Coordinator for the Division of Business and Economic Research. In 1973, he received his Ph.D. in regional science at the University of Pennsylvania and joined the University of New Orleans faculty. He served as Executive Assistant for Planning and Development to the mayor of New Orleans from 1978 to 1984, when he started a professional consulting firm specializing in urban and regional planning. He currently oversees and directs all planning projects and studies at the Mumphrey Group, Inc. He has published several journal articles, monographs, and reports, and he is a member of the American Institute of Certified Planners.

About the Contributors

Ronald Berkman is a Professor at Brooklyn College, a member of the doctoral faculty in political science, and the current University Dean for Academic Affairs. He received his Ph.D. from Princeton University, and he has taught at the Woodrow Wilson School of Public and International Affairs at Princeton, the University of California at Berkeley, New York University, and the University of Puerto Rico. He has written three books and numerous articles, and his recent publications include a chapter titled "The Re-Education of the American Working Class." He was the principal researcher for Mayor Dinkins's Urban Summit held in New York in November 1990 and has edited a volume based on the Summit.

Elise M. Bright completed her undergraduate work at the University of Arizona. After graduating magna cum laude in three years with a double major, she accepted employment as a social worker for the welfare department. The job piqued her interest in urban revitalization, and she decided to attend graduate school in city planning. She earned a master's degree from Harvard in 1975, followed by a doctorate from Texas A&M in 1980. She has worked in city and regional planning positions throughout the United States and has also conducted research in Europe. She has also been employed as a social worker, a newspaper editor, and a realtor. Her interest in urban revitalization has been enhanced by her personal experience as president of a small housing rehab/investment company. Since 1988 she has been on the faculty of the School of Urban and Public Affairs at the University of Texas at Arlington, where she specializes in teaching economic development, zoning, and environmental planning. In recent years she has conducted research into the effects of property assessment and tax collection methods on ownership in low-income neighborhoods and the unmet housing needs of nonnuclear households.

Richard L. Cole is Dean of the School of Urban and Public Affairs, University of Texas at Arlington. He received his Ph.D. from Purdue University in 1973. He came to the University of Texas at Arlington in 1980 from Yale University. Prior to that, he was Associate Professor of Political Science and Public and International Affairs at George Washington University. He is the author of eight books and more than 50 articles, primarily in the fields of urban affairs and urban policy analysis. His articles have appeared in *American Political Science Review, Public Administration Review, Political Science Quarterly, American Journal of Political Science, Western Political Quarterly,* and most of the regional journals in political science and urban politics. He is active in the profession, presenting papers at the annual meetings of various scholarly associations, serving as program chair of many scholarly meetings, and serving on the editorial boards of *Public Administration Review, Journal of Urban Affairs,* and other journals. He has served as President of the Southwest Political Science Association and as President of the American Society for Public Administration, North Texas Chapter, and has held offices in a variety of other professional associations. He has been recognized as Distinguished Alum by Purdue University and the University of North Texas.

David Imbroscio is Assistant Professor of Political Science at the University of Louisville. His recent work has appeared in the *Journal of Urban Affairs* and *Urban Affairs Quarterly.* He is currently writing a book on urban regime formation and alternative local economic development policy.

Mickey Lauria is Associate Professor and Director of the Division of Urban Research and Policy Studies in the College of Urban and Public Affairs at the University of New Orleans. He has published numerous articles on the politics of economic development. His current research interests focus on comparative urban redevelopment strategies in the United Kingdom, Eastern Europe, and the United States.

Anthony J. Maniscalco received his B.A. in political science and philosophy from Brooklyn College in 1989 and is currently working toward a doctorate in political science at the Graduate Center of the City University of New York. He teaches courses in American politics, political theory, and urban politics at two units of CUNY, John Jay College of Criminal Justice and Brooklyn College. He has been a staff researcher at the Central Office of the City University for the past three years.

Jeffrey H. Milgroom is a recent graduate of the Master of City Planning Program at the Georgia Institute of Technology, where he specialized in urban and economic development. He has a bachelor's degree, magna cum laude, from Clark University in geography, and is a member of Phi Beta Kappa. Prior to undertaking his graduate studies, he conducted market analysis for a New England consulting firm and served on the staff of a community development corporation.

Arthur C. Nelson is Professor of City Planning, Public Policy, and International Affairs at the Georgia Institute of Technology. He is widely published in the areas of regional development planning, urban form, resource land preservation, infrastructure planning and finance, growth management, and urban revitalization. He serves as an editor of the *Journal of the American Planning Association* and as associate editor of the *Journal of Urban Affairs*. His clients have included the Federal National Mortgage Association, National Academy of Sciences, U.S. Department of Transportation, U.S. Department of Commerce, U.S. Department of Housing and Urban Development, the Department of the Environment for the United Kingdom, and numerous regional, state, and local government associations. He has also consulted for corporations, development industry associations, and public policy associations. He earned his doctorate degree in urban studies from Portland State University in regional science and regional planning. Prior to his appointment at Georgia Tech, he served on the faculties of Kansas State University and the University of New Orleans.

Marion Orr is Assistant Professor of Political Science at Duke University. He earned his Ph.D. from the University of Maryland, College Park, and he has held fellowships from the Brookings Institution, the Ford Foundation, and the University of California, Berkeley. His research interests are in the areas of urban politics and public policy and African American politics. His research has appeared in the *Journal of Urban Affairs, Urban Affairs Quarterly, Urban Review,* and other scholarly publications. He is currently completing a study comparing the role of black mayors in Baltimore and Detroit in shaping the local agendas for human capital development. He is also conducting research on the politics of school reform in Baltimore.

Edward T. Rogowsky is Professor of Political Science at Brooklyn College of the City University of New York. He directs the college's off-campus Graduate Center for Worker Education and undergraduate public service internship programs. He has written several books and numerous articles on urban affairs, including, most recently, coauthorship of *Changing New York City Politics* (1991). He was appointed a member of the New York City Planning Commission in 1990 and continues to serve on that body. He also hosts a weekly urban affairs program, *MetroView,* broadcast citywide on CUNY-TV.

Timothy Ross is a Ph.D. student in the Department of Government and Politics at the University of Maryland, College Park. His current research focuses on community organizations and their impact on urban revitalization, and he has a special interest in the politics of New York City. He has recently coauthored a piece with Clarence Stone titled *Reinventing Local Governance: Competing Trends in the American Experience.*

Alex Schwartz is Assistant Professor at the Graduate School of Management and Urban Policy at the New School for Social Research, and Senior Research Associate at the New School's Community Development Research Center. He holds a Ph.D. in urban planning and policy development from Rutgers University. His research interests include housing and community development, service industries, and urban economic development. His most recent research projects include national studies of the management of housing owned by community development corporations and other nonprofit organizations and of the economic linkages between central cities and their suburbs. His articles have appeared in the *Journal of the American Planning Association, Urban Affairs Quarterly, Urban Geography,* and other journals.

Clarence Stone is Professor of Government and Politics at the University of Maryland, College Park. His research interests include the areas of urban education and related human investment policies, the restructuring of local governance, and the politics of social reform. Currently he is directing a multicity study titled "Civic Capacity and Urban Education," a 3-year project funded by the National Science Foundation. His most recent book is *Regime Politics: Governing Atlanta, 1946-1988.*

Elizabeth Strom is a doctoral candidate in political science at the Graduate Center of the City University of New York. Her dissertation research, which examines the politics of central-city development in Berlin since German unification, has been supported by grants from the Fulbright Program, the Social Sciences Research Council, the Germanistic Society of America, the McCune Foundation, and Arthur Schlesinger Jr. She is the holder of a master's degree in city planning from the Massachusetts Institute of Technology and a master's degree in political science from the CUNY Graduate Center. As part of CUNY's Wagner Institute, she has participated in a research project on the use of Community Development Block Grant funds in New York City, and she has investigated the use of public-private partnerships for urban development in five American cities for the German Institute of Urban Studies in Cologne. She has recently completed a study of economic regeneration in the German city of Rostock. Her current research focuses on cross-national comparisons of urban development.

Robert K. Whelan is Professor and Associate Dean in the College of Urban and Public Affairs at the University of New Orleans. He is coauthor of *Urban Policy and Politics in a Bureaucratic Age* (2nd ed., 1986) and has authored or coauthored many articles and papers, including "Urban Regimes and Racial Politics in New Orleans" (*Journal of Urban Affairs,* 1994). His research interests center on urban economic development and local government administration in the United States and Canada.

Sherman M. Wyman is Associate Professor in the School of Urban and Public Affairs and Coordinator of the M.P.A. Program at the University of Texas at Arlington. Previously, he was a faculty member and director of the City Management Graduate Program at the University of Kansas and Head of the Bureau of Governmental Research and Service at Denver University. He holds a bachelor's degree from Stanford University, an M.P.A. from Syracuse and a Ph.D. in public administration from the University of Southern California. A Fulbright scholar in comparative local governments in Norway in 1960-1961, he has held a variety of urban management and planning posts in addition to domestic and international consulting assignments. His research interests and publications focus on regional and urban administration and development, organizational change, and local economic development.

Alma H. Young is Professor of Urban Studies and Planning, College of Urban and Public Affairs, University of New Orleans. She has been actively involved in planning initiatives and other civic activities in the city of New Orleans. Her research interests include the political economy of urban development, issues of gender in the urban environment, and Caribbean political development. She is currently working on a manuscript on the role of ethnicity in the political development of Belize, in Central America.